Ross

ANNOTATED CHECKLIST OF THE BIRDS OF ONTARIO

Second Edition
Revised and Expanded

ROM
ROYAL ONTARIO MUSEUM
TORONTO

LIFE SCIENCES
MISCELLANEOUS PUBLICATIONS

ROYAL ONTARIO MUSEUM
PUBLICATIONS IN LIFE SCIENCES

The Royal Ontario Museum publishes three series in the Life Sciences:
CONTRIBUTIONS: a numbered series of original scientific publications.
OCCASIONAL PAPERS: a numbered series of original scientific publications, primarily short and of taxonomic significance.
MISCELLANEOUS PUBLICATIONS: an unnumbered series on a variety of subjects.

All manuscripts considered for publication are subject to the scrutiny and editorial policies of the Life Sciences Editorial Board, and to independent refereeing by two or more persons, other than Museum staff, who are authorities in the particular field involved.

LIFE SCIENCES EDITORIAL BOARD
Senior editor: J. H. McAndrews
Editor: M. D. Engstrom
Editor: G. B. Wiggins
External editor: C. S. Churcher

Manuscript editor: D. C. Darling
Production editor: K. K. Mototsune

Ross D. James is associate curator, Department of Ornithology, Royal Ontario Museum.

The Royal Ontario Museum is an agency of the Ontario Ministry of Culture and Communications.

Canadian Cataloguing in Publication Data

James, Ross D., 1943–
  Annotated checklist of the birds of Ontario

(Life sciences miscellaneous publications)
2nd ed., rev. and expanded.
Includes bibliographical references and index.
ISBN 0-88854-394-8

1. Birds - Ontario. I. Royal Ontario Museum.
II. Title. III. Series.

QL685.5.O6J3   1991   598.29713   C91-093546-7

Publication date: 17 June 1991
ISBN 0-88854-394-8
ISSN 0082-5093
© Royal Ontario Museum
100 Queen's Park, Toronto, Canada M5S 2C6

Typesetting by Q Composition Inc.
Printed and bound in Canada at University of Toronto Press.

Cover Illustration: Peregrine Falcon, *Falco peregrinus,* by Ross D. James.

CONTENTS

Introduction 5
    Treatment of rarities 6
    Species names 7
    Breeding status 8
    Status and distribution 8
    Dates of occurrence 12
    Egg dates 12
    Subspecies 12
    Abbreviations 13
Checklist of Species 14
Acknowledgements 86
Appendices 87
    Appendix I: Possible Escapees 87
    Appendix II: Comments on Subspecific Variation 88
Literature Cited 97
Additional References 109
Index to Common and Scientific Names 114

INTRODUCTION

This annotated list presents a summary of the species that have been recorded in the province of Ontario, with information about their status and breeding, distribution and abundance at various seasons, dates of occurrence, breeding seasons as indicated by egg dates, and subspecific relationships. Records and information to the end of 1989 have been included, but only a few 1990 records could be added.

It is now more than a decade since the forerunner of this checklist was published (James, McLaren, and Barlow, 1976). In the intervening years there has been a tremendous amount of study undertaken on Ontario birds: the publication of two volumes dealing with the breeding distribution and nesting habits of Ontario birds (Peck and James, 1983, 1987); five years of intensive fieldwork culminating in an atlas of the breeding birds of Ontario (Cadman, Eagles, and Helleiner, 1987); the appearance of a large summary volume on birds of Ontario (Speirs, 1985); and the production or revision of many regional works (Baxter, 1985; Bradstreet and McCracken, 1978; Brewer, 1977; Fazio, Shepherd, and Woodrow, 1985; Francis, 1984, 1985; Kelley, 1978; McCracken, 1987; McCracken, Bradstreet, and Holroyd, 1981; McRae, 1982; Mills, 1981; Nicholson, 1981; Sadler, 1983; Sandilands, 1984; Skeel and Bondrup-Nielsen, 1978; Sprague and Weir, 1984; Weir, 1989b; Weir and Quilliam, 1980; Wormington, 1982).

There has been, in addition, a revised edition of *The Birds of Canada* (Godfrey, 1986), and the American Ornithologists' Union has published a new *Check-list of North American Birds* (AOU, 1983). The Ontario Bird Records Committee (OBRC) replaced the Ontario Ornithological Records Committee, and this has resulted in some significant changes in the handling of bird records and has initiated a review of past records of various sorts. Numerous species have overwintered successfully farther north than ever before, owing to the ever expanding presence of bird feeders, the warmth of buildings, and climatic changes. There has also been a continuing increase in the number of field observers and their ability to identify birds. Their observations in the seasonal summaries of *American Birds* provide increasingly important sources of new information about Ontario birds.

These many factors have produced an incredible number of changes and additions to the information about the birds of Ontario. Much of what was published in the previous checklist is now known to have been incomplete and, in some instances, incorrect. Practically no species account has been exempt from at least a minor revision, and numerous species have been added to the list. Although several other books are available about the birds of Ontario, a checklist like this still has a role to play, providing a convenient summary of information for a particular time period, and from many sources not found in any other single work. Undoubtedly this compilation still contains some deficiencies and there will be further refinements as the understanding of birds continues to grow. But a much improved checklist should provide a summary basis for continued growth and a reference against future changes.

The present compilation considers 442 species as having been reported in Ontario with adequate documentation, and an additional 12 species for which some further confirmation is desirable. Of these, 285 species are considered to have bred at least once in the province, and an additional 7 have probably bred, although material evidence is lacking. Three forms considered as species (American Black Duck, Carolina Chickadee, and Hoary Redpoll) may not warrant species status—see Appendix II.

Treatment of rarities

At this point in the history of Ontario ornithology, specimen collecting is not carried out as it was in the past, when rarities were shot as a matter of course. Most records of unusual occurrences in the province are now in the form of photographs or of written descriptions without specimens. But the information available to observers to help identify birds in the field has never been better, and vagrant birds are usually the objects of scrutiny by numerous observers. There is also an established bird records committee to review records and publish an annual report. The present records committee, of which I have been a member, is also prepared to accept, in some circumstances, a first record for the province in the absence of a specimen or even photographic evidence. It seems unrealistic to continue to reject a species as occurring in Ontario, because there is no specimen or photograph, if there is good written information, especially where multiple observations of a single event or several observations of repeated events have been preserved.

On the other hand, it is not possible to agree completely with any records committee that has been or is now operating. In the past, any first record was automatically rejected without a specimen or photograph. Some of these records have not been reconsidered by the OBRC, which might now be accepted. Moreover, many old records have no written descriptions, but may represent perfectly good sightings by competent observers. There was no records committee prior to 1970, and no descriptions were preserved. We have no way now to evaluate such sightings objectively; we can only consider the qualifications of the observer and the degree of difficulty of the identification. There is also the problem of specimens that are now missing. Were they correctly identified at the time or, as has sometimes happened, were identities based on inadequate knowledge?

We are likewise faced with the problem of escaped captives. For many occurrences, we can never be sure whether a bird had human assistance, either deliberate or inadvertent, in reaching Ontario. We have only probabilities and human judgement, and not everyone is likely to agree. There are also many sightings of birds for which no written report is ever submitted to the records committee. These may be accurate, but no assessment is possible.

Use of the term "hypothetical" has generated and continues to generate considerable controversy, and what should be included in such a category is a matter of personal opinion. Hence this term has been removed from the checklist. I have placed in square brackets the entry on any species that I feel requires further documentation. But, whether the species is so designated or not, I have attempted to indicate what the available documentation is for vagrant species, and have indicated where controversy exists. Where I have expressed a personal opinion, others are free to disagree. The avaiiable information, whatever it is, will have to stand on its own. All submissions to the OBRC are housed in the Royal Ontario Museum (ROM) and are available for further examination.

An entry has been included for any unusual species for which there is a previously published reference (except for clearly rejected records). Thus, the reader has a guide to the acceptability of such references, especially where this had not been clear. Individual records of occurrence are listed only where five or fewer are known from the province. No mention has been made of the many obvious escaped captives that are frequently reported. A number of probable escapees, but possibly wild birds, are mentioned in Appendix I following the species accounts. These are species for which there are few or no North American records beyond the species' normal range or near Ontario. There are a number of statements of unusual sightings in museum files for which there are no descriptions or other supporting materials that might be used to evaluate them. Such unpublished sightings have not been mentioned.

No attempt has been made to give a complete listing of hybrids. Brewster's and Lawrence's warblers are the only two distinctly plumaged and named hybrids of annual occurrence and are mentioned with the parental types.

Species names

Common and scientific names, for the most part, follow the American Ornithologists' Union (AOU) *Check-list of North American Birds* (1983) and its 35th and 37th supplements (AOU, 1985, 1989). Differences occur only in relation to Thayer's/Iceland gulls (see Appendix II). I have included a four-letter code abbreviation for each species. The codes used, with few exceptions, are those used by the United States Fish and Wildlife Service for their North American bird banding manual, revised in July 1988. Where the service has not provided codes, I have followed its rules for the establishment of codes. Exceptions to its list include species for which it uses outdated nomenclature (Tundra Swan, Barn Owl, Northern Rough-winged Swallow, American Pipit), and where it provides separate codes for two or more populations or colour morphs of the same species (Green-winged Teal, Snow Goose, Brant, Northern Flicker, Tufted Titmouse, Yellow-rumped Warbler, Palm Warbler, Dark-eyed Junco, Northern Oriole, and Rosy Finch).

Breeding status
An asterisk following the name indicates that the species has nested in the province. An asterisk in brackets indicates that breeding is suspected or known, but has not been adequately documented with specimens, photographs, or written descriptions.

Status and distribution
The distribution of any species is seldom precise. The numbers of birds will decline as the periphery of a range is approached, and the extent to which those at the edges will wander may vary from year to year. In order to delineate range, the names of communities, current political subdivisions, or water bodies are used to give an approximate limit. The locations or physical features used in the text are given in Figures 1 and 2.

Any bird whose range is described as being in the "south" has a distribution partially or wholly south of 47 degrees north latitude, which crosses Ontario just to the north of Sault Ste. Marie and Sudbury (Figs. 1 and 2). This line of north–south division is somewhat farther north than that given in James, McLaren, and Barlow (1976), but includes areas cleared of forests lying immediately north of the North Channel of Lake Huron. These lands are more similar to southern Ontario than to the boreal forests, and are included in the range of several more southern species (Coady, 1988). Conversely, birds with ranges in the "north" occur north of this latitude. The "west" of the province always refers to the northern part of Ontario, west of the longitude of Lake Nipigon. The "north coast" includes the coasts of Hudson and James bays. Although the Winisk town site has been abandoned, the name has been retained on the map. This locality appears in many earlier ornithological publications, and numerous records come from there. For the extents of the Deciduous Forest, Great Lakes/St. Lawrence Forest, Boreal Forest, and Tundra regions, and of the Hudson Bay Lowland, the Canadian Shield, and the farming areas of the Clay Belt, please refer to Figures 1 and 2.

In defining relative abundance and frequency of occurrence at various seasons, it is difficult to do more than give a general impression. Numbers of birds vary from the centre to the periphery of the range, as the seasons change, or even from year to year at the same locality, and will reach a maximum only in areas of suitable habitat. Nevertheless, I have attempted to provide information about the frequency, abundance, and seasonal occurrence of species that could be expected at the appropriate time and place.

There have been many systems used to define abundance and frequency; none seems to be ideal and everyone seems to prefer a different version. In general, the simpler the system, the more easily it is understood. A simpler system may be less precise; however, given that Ontario is large in size and in habitat diversity, that we have less than perfect information, and that birds tend not to follow strict rules, the terms given below seem reasonable.

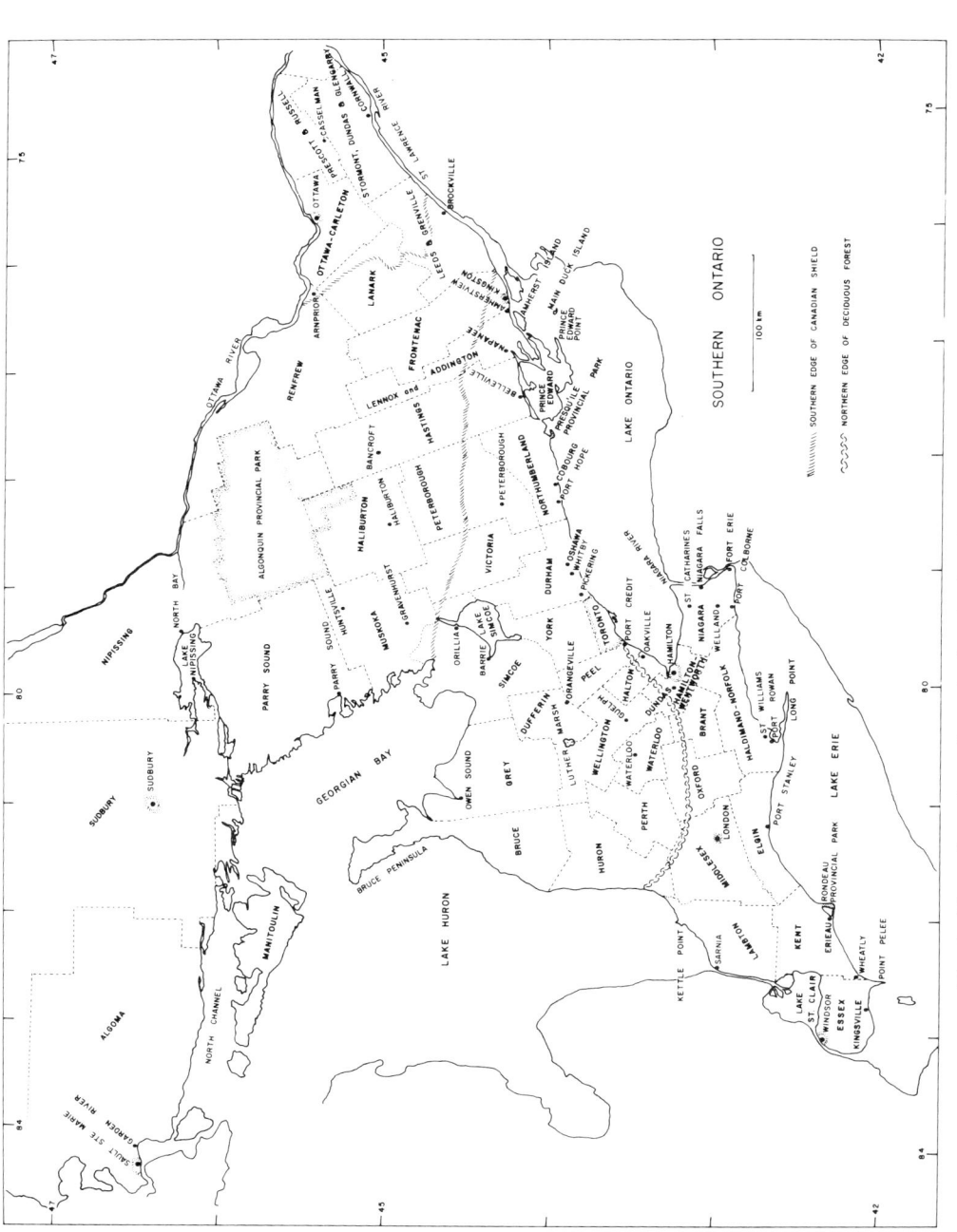

Fig. 1. Map of southern Ontario showing localities mentioned in the text.

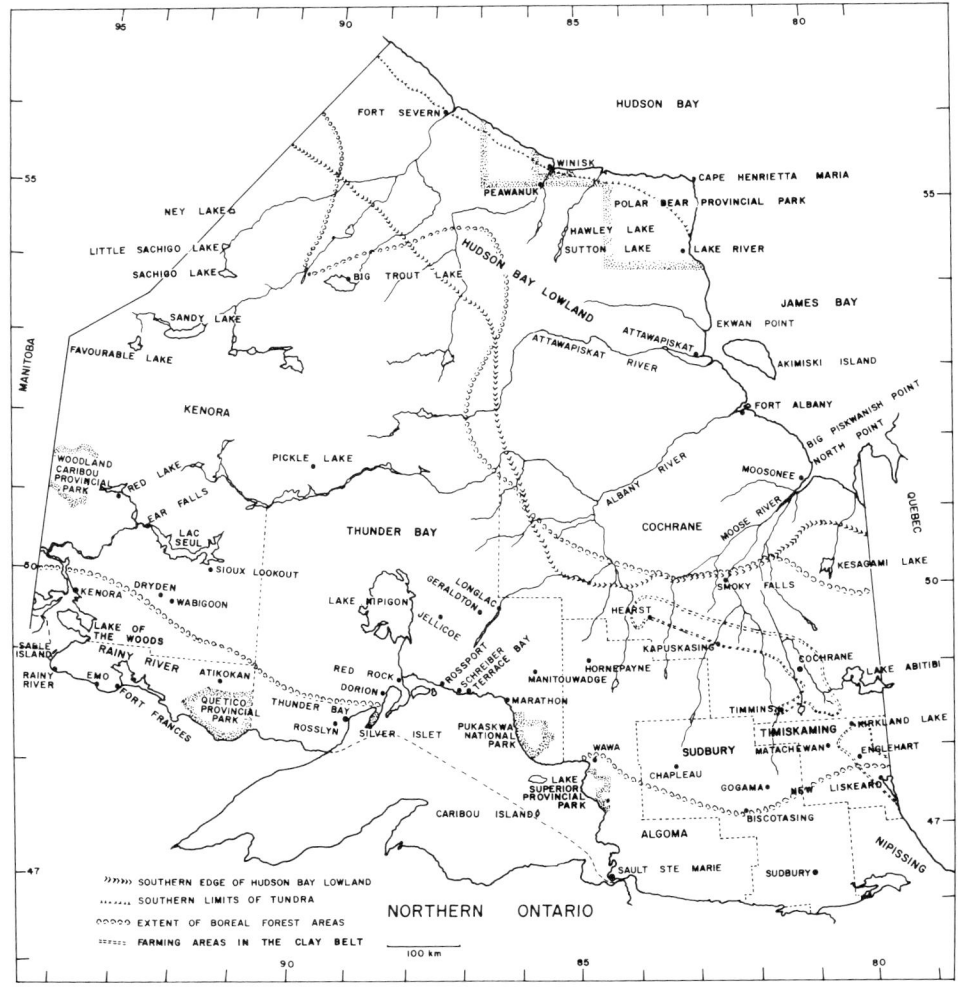

Fig. 2. Map of northern Ontario showing localities mentioned in the text.

In this volume the system is much the same as in the previous edition (James, McLaren, and Barlow, 1976) and roughly comparable terms are used.

Frequency related to annual occurrence
  regular expected every year (usually not specified since most birds fit this category)
  occasional not expected every year, but to be expected most years and usually at least once a decade
  vagrant has occurred, but five or fewer records exist, and the probability of recurrence is very low (usually less than once a decade)

Relative abundance   the usual numbers that could be seen during a specified season in the appropriate habitat away from the periphery of the range. An abundance term in parentheses is a condition much less likely to be encountered.
- **abundant**   more than 500 could be seen in a day
- **common**   can always be found (well distributed); more than 25 a day is usual, but seldom more than 500
- **uncommon**   can usually be found, but numbers seen are likely to be small (6 to 25 a day)
- **rare**   usually seen singly (seldom more than 5) and difficult to find on any particular outing (unless a specific location is known)

Seasonal occurrence
- **resident**   passing one or more seasons in the province. Winter residents (as opposed to permanent residents normally expected through winter) are thought to have spent an entire winter at least once, even though many others of their kind occurring in winter may be only late departures or early arrivals.
- **migrant**   regularly passing through at least part of the province, coming to breed or leaving for the winter
- **transient**   regularly migrating through or into the province, but not breeding here
- **straggler**   irregularly wandering in or into, or remaining behind in, the province, usually in small numbers, and sometimes at an unusual time or place
- **irruptive**   moving within or into Ontario, sometimes in large numbers, in some years only; usually moving southward in autumn from or through normal range
- **erratic**   wandering widely, usually in winter flocks; may or may not be seen in any particular area in any given year

Several additional terms are used. **Extinct** species are those no longer living anywhere on earth. **Extirpated** species are not extinct, but no longer reside within the geographic area under consideration. **Introduced** species are those that have had human assistance in establishing themselves on the North American continent. Introductions that have not become successfully established are not mentioned.

For occasional species, I have attempted to give an impression of the frequency of occurrence by listing the date that the species was first reported in Ontario, and the approximate number of reports in the past decade (or other appropriate time period) or the date of last occurrence. The number of reports is usually approximate since many of the sightings remain undocumented in written or other form.

Dates of occurrence

For most species, other than permanent residents, dates that indicate the time of year when the species has been recorded in the province are provided. Four dates are given for most migratory species. The first and last dates, in parentheses, represent the extremes of early and late occurrences, respectively. The middle two dates indicate the time period during which nearly normal numbers are usually found in some part of the province. In any year, there will be some variation in the time of arrival and departure, but the dates of usual occurrence indicate times when, even in exceptional circumstances, there almost certainly will be birds present. These dates apply largely to the extreme south of the province where the first and last migrants are normally expected. Where species are largely migratory, but considerable numbers remain in Ontario throughout the winter, only dates of greatest abundance (principal dates) are given. Since there is annual variation in the occurrence of any species, precise dates are not used (except for records of vagrants). The "early" part of the month refers to the first ten days, "mid" from the 11th to the 20th, and "late" the latter third of the month from the 21st on.

I have not attempted a rigid definition of seasons by specific calendar months. Various species migrate, breed, and winter at different times of the year, and even within a species some individuals may be migrating while others are breeding, etc. Thus, even though there might be records of a species in December, I would not necessarily consider it a species occurring in winter, but probably only a late migrant. A bird would have to be found on one or more occasions or through the January–February period to be a winter species. Likewise a bird found in June would not necessarily be a summering bird, but could be a late northward-bound migrant. The behaviour or intent of the bird or birds during the appropriate season for that species may be more important than a specific calendar date.

Egg dates

These dates give an indication of the nesting season for each species. More specific details are available in Peck and James (1983, 1987). If egg dates are unknown, breeding has been established by the presence of flightless young, with the exception of the Black Guillemot, whose hatched eggshells were found.

Subspecies

A review of subspecific variation occurring in Ontario is undertaken here. No attempt is made to identify new subspecies, but only to verify or comment on the distribution or occurrence of forms already named. The extensive ROM collections from Ontario were used to make most determinations. Collections of the Canadian Museum of Nature also were consulted for a number of species. In a few instances there is inadequate material available to do more than concur with the AOU checklist designations (AOU,

1957). Where there is only a single form occurring in North America or anywhere near the province, the subspecific designations of the AOU (1957) are used without attempting to verify them. No designations are given for species whose occurrence is based solely on sight records or photographic records, unless subspecific differences are obvious.

Boundaries of subspecies' ranges are not precise. Birds tend to move widely and intergrade extensively within contiguous populations. Subspecific differences are often subtle, and wide zones of intergradation may be encountered before discrete populations can be identified. Especially in southern Ontario (and neighbouring states), human activity has greatly disturbed the environment and may have altered migration and breeding distributions, breaking down subspecies populations that were formerly more discrete.

Where differences are evident from the AOU checklist (AOU, 1957) or from James, McLaren, and Barlow (1976), with respect to a subspecies' (or species') occurrence or to their distributions in the province, more detailed comments are to be found in Appendix II, outlining the reasons for those differences.

Abbreviations

The following abbreviations have been used in this volume: AOU (American Ornithologists' Union); BSNS (Buffalo Society of Natural Sciences); DM (District Municipality); NMC (National Museum of Canada–Canadian Museum of Nature); OBRC (Ontario Bird Records Committee); RM (Regional Municipality); ROM (Royal Ontario Museum); ROM PR (ROM photo-record); UMMZ (University of Michigan Museum of Zoology); USNM (United States National Museum).

CHECKLIST OF SPECIES

Red-throated Loon, *Gavia stellata* **(Pontoppidan)*** RTLO
Rare (to locally uncommon) summer resident on the Hudson Bay coast; occasional, rare straggler and historical breeder on Lake Superior. Occasional, rare straggler in summer on the lower Great Lakes (nonbreeding). Rare straggler in winter in the south (north to Ottawa). Rare (to locally common) migrant. Principal dates in province: early April to mid November. Egg dates: 30 June to 3 August.

Pacific Loon, *Gavia pacifica* **(Lawrence)*** PALO
Uncommon local summer resident along the coast of Hudson Bay. Rare migrant elsewhere. Dates in province: (mid April) mid May to early November (late December). Egg dates: 23 June to 14 July.

Common Loon, *Gavia immer* **(Brünnich)*** COLO
Common summer resident throughout most of the province, rare and irregular, with few exceptions, south of the Canadian Shield to Lake Erie. Rare winter resident in the south; rare straggler in the north. Common (to locally abundant) migrant. Principal dates in province: early April to late October. Egg dates: 11 May to 25 August.

Yellow-billed Loon, *Gavia adamsii* **(Gray)** YBLO
Occasional, rare straggler throughout the province; recorded in May, June, and December. First reported in 1957 (Gunn, 1957; Beardslee and Mitchell, 1965); one record in the past decade.

Pied-billed Grebe, *Podilymbus podiceps* **(Linnaeus)*** PBGR
Rare to locally uncommon summer resident, mainly south of the Canadian Shield; becoming rare north to Big Trout Lake and southern James Bay. Occasional, rare winter resident in the lower Great Lakes region. Uncommon to locally common migrant. Principal dates in province: early April to late October. Egg dates: 3 May to 22 August.
   P. p. podiceps.

Horned Grebe, *Podiceps auritus* **(Linnaeus)*** HOGR
Rare (and occasional?) summer resident in the extreme northwest (Fort Severn); occasional, rare, possible breeder in the south (mainly Lakes Ontario, Erie, and St. Clair). One substantiated breeding record since 1938. Occasional, rare straggler in winter in the lower Great Lakes region. Uncommon (to locally common) transient and migrant. Principal dates in province: early April to late November. Egg dates: 16 May to 1 July.
   P. a. cornutus.

Red-necked Grebe, *Podiceps grisegena* **(Boddaert)***　　　　　**RNGR**
Rare to uncommon summer resident in a few isolated colonies, north to Big Trout Lake and Cochrane. Most likely to be found in the west (Thunder Bay to Red Lake region). Rare straggler in winter in the lower Great Lakes region; vagrant on Lake Superior. Uncommon to locally common migrant. Principal dates in province: mid April to early November. Egg dates: 15 May to 17 September.
　P. g. holbollii.

Eared Grebe, *Podiceps nigricollis* **Brehm**　　　　　**EAGR**
Vagrant in summer. Occasional, rare straggler in winter in the lower Great Lakes region. Rare transient in the south, also in Rainy River area. Principal dates in province: mid April to May; September to November.
　P. n. californicus.

Western Grebe, *Aechmophorus occidentalis* **(Lawrence)**　　　　　**WEGR**
Occasional, rare transient in the south and at Thunder Bay, mainly in spring and autumn, but also in summer. Reported since the 1880s (specimen); at least two reports in the past decade. Principal dates in province: April to May; mid September to mid November.

Northern Fulmar, *Fulmarus glacialis* **(Linnaeus)**　　　　　**NOFU**
Occasional, rare straggler in spring and autumn in the south (two records between 1978 and 1988); rare (to uncommon) transient in autumn on the north coast; vagrant elsewhere in the north (Manitouwadge). Dates in province: early to mid May; late October to mid December.
　F. g. minor usual straggler and northern interior vagrant.
　F. g. glacialis vagrant.

Black-capped Petrel, *Pterodroma hasitata* **(Kuhl)**　　　　　**BCPE**
Vagrant; three specimens: Toronto, 30 October 1893 (Brown, 1894); Oakville, "about 1893" (Fleming, 1906); Port Colborne, 21 August 1955 (Baillie, 1955).

Audubon's Shearwater, *Puffinus lherminieri* **Lesson**　　　　　**AUSH**
Vagrant; one specimen: (NMC 62529) Almonte, Lanark County, 8 September 1975 (Godfrey, 1976).
　P. l. loyemilleri (Godfrey, 1986).

Wilson's Storm-Petrel, *Oceanites oceanicus* **(Kuhl)**　　　　　**WISP**
Vagrant; two specimens: (ROM 34288) Muskoka DM, spring 1897 (Fleming, 1901); (ROM 73077) Niagara RM, 14 August 1955 (Baillie, 1955).
　O. o. oceanicus.

Leach's Storm-Petrel, *Oceanodroma leucorhoa* **(Vieillot)**   **LHSP**
Vagrant; two specimens: (ROM 33400) Cornwall, 19 July 1939 (Toner, 1940); (ROM 73086) near Kingston, 16 August 1955 (Baillie, 1955). One sight record: Attawapiskat, 8 October 1981 (James, 1983).
   O. l. leucorhoa.

Band-rumped Storm-Petrel, *Oceanodroma castro* **(Harcourt)**   **BSTP**
Vagrant; one specimen: (NMC 25688) Ottawa, 28 August 1933 (Taverner, 1934).
   O. c. castro.

Northern Gannet, *Morus bassanus* **(Linnaeus)**   **NOGA**
Occasional, rare straggler throughout, mainly in autumn in the Great Lakes area. Dates in province: mid October to early February (also mid May and late June).

American White Pelican, *Pelecanus erythrorhynchos* **Gmelin***   **AWPE**
Locally abundant summer resident at Lake of the Woods; occasional, rare straggler elsewhere in the north in spring and summer. Rare straggler in southern Ontario in spring, summer, and autumn. Dates in province: (early February) late April to early October (late December). Egg dates: 8 May to 27 July.

Brown Pelican, *Pelecanus occidentalis* **Linnaeus**   **BRPE**
Vagrant; one photographic record: (ROM PR 22–25) Wavery Beach, Niagara RM, 25 September 1971 (Goodwin, 1972b).

Great Cormorant, *Phalacrocorax carbo* **(Linnaeus)**   **GRCO**
Occasional, rare straggler on the Great Lakes and Ottawa River, mainly during the winter months. One specimen: (ROM 34423) Toronto, 21 November 1896 (Fleming, 1900). One photographic and at least seven sight records since 1978. Dates in province: (August) December to March (late May).
   P. c. carbo.

Double-crested Cormorant, *Phalacrocorax auritus* **(Lesson)***   **DCCO**
Locally common (to abundant) summer resident, mainly on the Great Lakes, Lake of the Woods, Lake Abitibi, and Lake Nipigon; but also north to Little Sachigo Lake and formerly on James Bay. Vagrant north to Fort Severn (Manning, 1952). Occasional, rare winter resident on the southern Great Lakes. Rare to locally abundant migrant. Principal dates in province: early April to late October. Egg dates: 27 April to 30 August.
   P. a. auritus.

Anhinga, *Anhinga anhinga* **(Linnaeus)** ANHI
Vagrant; one specimen: (now missing) West Lake, Prince Edward County, 7 September 1904 (Snyder, 1941), seems reasonable based on Snyder's comments. A specimen (UMMZ 91960) from Garden River, Algoma District, from 1881 (Van Tyne, 1950) has been rejected (see Payne, 1983) as a valid record.
 [*A. a. leucogaster.*]

Magnificent Frigatebird, *Fregata magnificens* **Mathews** MAFR
Vagrant; one photographic record: Sarnia, 28 September 1988 (Weir, 1989a; Wormington and Curry, 1990). A sight record mentioned by Baillie (1950b) has been rejected by OBRC.
 [*F. m. rothschildi.*]

American Bittern, *Botaurus lentiginosus* **(Rackett)*** AMBI
Rare to uncommon summer resident throughout the province. Occasional, rare straggler in winter in the south. Rare migrant. Principal dates in province: late April to late September. Egg dates: 5 May to 10 August.

Least Bittern, *Ixobrychus exilis* **(Gmelin)*** LEBI
Rare to locally common summer resident in the south (north to Sault Ste. Marie and North Bay); possibly also in Lake of the Woods area. Rare migrant; wanders north to Atikokan and Thunder Bay. Dates in province: (mid April) early May to mid September (early December). Egg dates: 15 May to 2 August.
 I. e. exilis.

Great Blue Heron, *Ardea herodias* **Linnaeus*** GTBH
Common summer resident in the south, uncommon in the north to Sandy Lake and Moosonee; wandering occasionally north to Hudson Bay. Rare winter resident in the south. Common migrant. Principal dates in province: early April to mid November. Egg dates: 24 April to 30 June.
 A. h. herodias.

Great Egret, *Casmerodius albus* **(Linnaeus)*** GREG
Uncommon (to locally abundant) summer resident in Essex County, rare and widely scattered elsewhere in the south, north to southern Georgian Bay. Occasional, rare straggler north of breeding range in spring and late summer (usually in the south, but north to Fort Frances, northern Lake Superior, and Kapuskasing). Population currently increasing. Rare (to locally common) migrant. Dates in province: (early March) mid April to mid September (late December). Egg dates: 11 May to 24 June.
 C. a. egretta.

Snowy Egret, *Egretta thula* **(Molina)*** SNEG
One nest record at Hamilton in 1986 (Curry and Bryant, 1987). Usually rare straggler in the south (north to Bruce County and Kingston); vagrant in the north (Rainy River, Thunder Bay, and Attawapiskat). Dates in province: (early April) early May to mid September (late October). Egg dates: unknown.
 E. t. thula.

Little Blue Heron, *Egretta caerulea* **(Linnaeus)** LBHE
Rare straggler in the south, mainly in spring, with sightings north to Sudbury; also at Thunder Bay and Winisk in the north. Dates in province: (late March) early May to mid October (mid December).
 E. c. caerulea.

Tricolored Heron, *Egretta tricolor* **(Müller)** TRHE
Rare straggler in the south (north to Simcoe and Prince Edward counties), and at Marathon in the north, in spring and summer. Dates in province: early April to early September.
 [*E. t. ruficollis.*]

Cattle Egret, *Bubulcus ibis* **(Linnaeus)*** CAEG
Occasional, rare local summer resident in the south (north to Kingston). Rare straggler in the rest of the province, north to Pickle Lake and Attawapiskat. Rare (to locally common) migrant. Dates in province: (early April) early May to late September (early December). Egg dates: 4 June to 14 July.
 [*B. i. ibis.*]

Green-backed Heron, *Butorides striatus* **(Linnaeus)*** GNBH
Uncommon (to locally common) summer resident in the south (north to Sault Ste. Marie and North Bay); occasional, rare in the north (to Rainy River, Thunder Bay, and Marathon). Uncommon (to locally common) migrant. Dates in province: (early April) mid April to early October (late December). Egg dates: 4 May to 25 July.
 B. s. virescens.

Black-crowned Night-Heron, *Nycticorax nycticorax* **(Linnaeus)*** BCNH
Locally common summer resident in the south (mainly south of the Canadian Shield), with reports north to northern Nipissing District (Peck and James, 1983). Occasional, rare straggler in the north in summer (to Thunder Bay and Cochrane). Occasional, rare straggler in winter. Uncommon migrant. Dates in province: (early March) early April to early October (mid January). Egg dates: 6 May to 9 September.
 N. n. hoactli.

Yellow-crowned Night-Heron, *Nyctanassa violacea* **(Linnaeus)**   **YCNH**
Occasional, rare straggler in the south (north to Ottawa), spring to autumn. Dates in province: (early April) early May to early October.
N. v. violacea.

White Ibis, *Eudocimus albus* **(Linnaeus)**   **WHIB**
Vagrant; two specimens: (now missing) Clayton, Lanark County, 13 October 1955 (Baillie, 1957); (ROM 113149) Long Point, summer 1965. Also a sight record at Point Pelee, 27 September 1970 (Goodwin, 1971a).

Glossy Ibis, *Plegadis falcinellus* **(Linnaeus)**   **GLIB**
Rare straggler in the south (north to Georgian Bay and Kingston), mainly in spring. Dates in province: (early April) early to late May (early November).
P. f. falcinellus.

[White-faced Ibis, *Plegadis chihi* **(Vieillot)**   **WFIB**
A September 1908 record cited by Beardslee and Mitchell (1965) concerns a second-hand report of a now missing specimen, which, even if correctly identified, was of questionable origin.]

Wood Stork, *Mycteria americana* **Linnaeus**   **WOST**
Occasional, rare straggler in the south (north to Algonquin Park), mainly in the late summer and autumn. First reported in 1892 (Fleming, 1913); no records since 1972. Dates in province: early May to early November.

[Greater Flamingo, *Phoenicopterus ruber* **Linnaeus**   **GREF**
Birds at Wheatly, 26 June 1968; at Dorion, 20 to 22 October 1978; in Prince Edward County, 23 September to 20 October 1978; and in Lambton County, 2 to 7 May 1979; all thought to be escapees. The origin of a bird in Niagara RM, 11 April to 24 May and 18 June 1968 (Sheppard, 1970) is unknown, but may also have been a former captive.]

Fulvous Whistling-Duck, *Dendrocygna bicolor* **(Vieillot)**   **FUWD**
Occasional, rare straggler in the south (north to Cornwall). Dates in province: early April to early December. First reported in 1960 (Barlow, 1966); at least five reports in the past decade.

Tundra Swan, *Cygnus columbianus* **(Ord)***   **TUSW**
Rare summer resident of the Hudson Bay coast; numerous records of non-breeding birds from southern Ontario and Lake Superior. Occasional, rare (to locally common) winter resident in the extreme south (also Owen Sound). Abundant migrant in the southwest; rare elsewhere. Principal dates in province: early March to early December. Egg dates: 6 to 25 June.
C. c. columbianus.

Trumpeter Swan, *Cygnus buccinator* **Richardson[*]**　　　　　　　　**TRUS**
Former migrant and probable former breeder (Lumsden, 1984b), extirpated by 1884. Current small-scale attempt to return them to Ontario, begun in 1982 (Lumsden, 1984a), has resulted in the possible presence of a few birds at any season on Lake Ontario, usually between Toronto and Whitby.

Mute Swan, *Cygnus olor* **(Gmelin)***　　　　　　　　　　　　　　**MUSW**
Uncommon permanent resident in the extreme south (Lakes Ontario and Erie); reported north to Lake Superior in summer, and to Haliburton in winter. Has been widely introduced and is maintained in many urban centres, but feral populations have been established at least since 1958 (Peck, 1966; Peck and James, 1983). Egg dates: 4 April to 29 June.

Greater White-fronted Goose, *Anser albifrons* **(Scopoli)**　　　　　**GWFG**
Vagrant in summer (nonbreeding) at Thunder Bay. Rare straggler in winter in the south. Rare (to locally uncommon) spring and autumn transient in the south and west. Dates in province: early February to early June; early October to mid December.
　A. a. frontalis usual transient.
　A. a. flavirostris vagrant (ROM 99246) on Moose River, 1 October 1966, and a few sight records of yellowish-billed birds.

Snow Goose, *Chen caerulescens* **(Linnaeus)***　　　　　　　　　　**SNGO**
Uncommon to locally abundant summer resident along the north coast, south to Albany River; straggler in the south (nonbreeding). Rare (to locally common) winter resident in the south (north to Toronto). Locally abundant migrant in the north; locally common (to abundant) in the south. Principal dates in province: mid March to late November. Egg dates: 2 to 22 June.
　C. c. caerulescens.

Ross' Goose, *Chen rossii* **(Cassin)***　　　　　　　　　　　　　　**ROGO**
Occasional, rare summer resident along the north coast. Rare migrant on the north coast. Dates in province: early May to mid October (early December). Egg dates: unknown.

Brant, *Branta bernicla* **(Linnaeus)[*]**　　　　　　　　　　　　　**BRAN**
Occasional, rare to uncommon straggler in summer on the north coast and in the south (nonbreeding). Unsubstantiated breeding reported at Sudbury in 1954 among birds which failed to migrate (Baillie, 1955), and on the St. Lawrence River among escapees or birds which failed to migrate (Peck and James, 1983; Lumsden, 1987a). Rare winter resident in the south (north to Prince Edward County). Abundant migrant along the James Bay coast and in the southeast; rare elsewhere. Principal dates in province: early May to mid June; early September to late November.
　B. b. hrota.

Barnacle Goose, *Branta leucopsis* **(Bechstein)** BRNG
Occasional, rare straggler in the south in spring and autumn. About a dozen sight records since first reported in 1955, some obviously escaped captives, but several probably wild birds. Dates in province: March to April; October to November.

Canada Goose, *Branta canadensis* **(Linnaeus)*** CAGO
Common summer resident in the Hudson Bay Lowland; becoming uncommon south to Sandy Lake and Cochrane. Introduced and feral birds now breed throughout most of southern Ontario, and in the Lake of the Woods, Thunder Bay, and probably along the eastern Lake Superior areas. Common to abundant winter resident along the lower Great Lakes. Abundant migrant. Egg dates: 29 March to 30 June.
- *B. c. interior* breeds mainly throughout the Hudson Bay Lowland and adjacent areas of northern Ontario, migrant elsewhere.
- *B. c. parvipes* probably a rare transient in the northwest.
- *B. c. hutchinsii* a rare transient, mainly on the north coast; occasional, rare winter resident in the south.
- *B. c. moffitti* breeds and winters in southern Ontario.

Wood Duck, *Aix sponsa* **(Linnaeus)*** WODU
Uncommon summer resident in the south, rare in the north to Lake of the Woods and Cochrane areas; straggler north to Sandy Lake and Moosonee. Rare winter resident in the south (north to Kingston). Common migrant. Principal dates in province: mid March to late October. Egg dates: 28 March to 18 July.

Green-winged Teal, *Anas crecca* **Linnaeus*** GWTE
Rare to uncommon summer resident across the province. Rare winter resident in the south (north to Sudbury); vagrant in the north (Thunder Bay). Common (to locally abundant) migrant. Principal dates in province: late March to mid November. Egg dates: 11 May to 6 July.
  A. c. carolinensis breeds.
  A. c. crecca vagrant February to April.

American Black Duck, *Anas rubripes* **Brewster*** ABDU
Uncommon summer resident throughout most of the province; becoming rare and local in southern agricultural areas and in the west. Populations currently experiencing declines. Common winter resident in the south (north to Sault Ste. Marie). Common (to locally abundant) migrant. Principal dates in province: early March to late November. Egg dates: 1 April to 18 July.

Mallard, *Anas platyrhynchos* **Linnaeus*** MALL
Common summer resident throughout the province. Common to locally abundant winter resident in the south (north to Sault Ste. Marie). Common migrant. Egg dates: 2 April to 20 July.
   A. p. platyrhynchos.

Northern Pintail, *Anas acuta* **Linnaeus*** NOPI
Common summer resident on the Hudson Bay coast and adjacent Lowland; uncommon south of the Canadian Shield; occasional, rare elsewhere. Uncommon winter resident in the south (north to Manitoulin Island). Common (to locally abundant) migrant. Principal dates in province: early March to mid November. Egg dates: 10 April to 30 June.

Blue-winged Teal, *Anas discors* **Linnaeus*** BWTE
Uncommon to locally common summer resident, mainly south of the Canadian Shield; becoming occasional, rare north to the Hudson Bay coast. Rare in winter in the south. Common (to locally abundant) migrant. Principal dates in province: late March to mid October. Egg dates: 4 May to 29 July.
   A. d. discors.

Cinnamon Teal, *Anas cyanoptera* **Vieillot*** CITE
Occasional, rare summer straggler, and unexpected breeder (one record: James, 1984a). Rare migrant in spring or autumn, mainly in the south, also in Lake of the Woods and Thunder Bay areas. Dates in province: mid April to mid November. Egg date: 24 June.
   A. c. septentrionalium.

Northern Shoveler, *Anas clypeata* **Linnaeus*** NSHO
Rare local summer resident, mainly in the south, but also to Lake of the Woods, Thunder Bay, and Cochrane in the north; rare local along the coasts of Hudson and James bays. Occasional, rare winter resident in the south. Uncommon (to locally common) migrant. Principal dates in province: late March to early November. Egg dates: 22 May to 23 June.

Gadwall, *Anas strepera* **Linnaeus*** GADW
Common local summer resident in the south; rare and scattered in the north to Hudson Bay. Populations have expanded rapidly in the last decade. Common winter resident in the south. Common migrant. Principal dates in province: late March to early November. Egg dates: 15 April to 25 July.

Eurasian Wigeon, *Anas penelope* **Linnaeus** EUWI
Vagrant in summer. Rare transient in the south (north to Sudbury); also to Thunder Bay, Marathon, and James Bay in the north. Dates in province: April and May (June and July); September to December.

American Wigeon, *Anas americana* **Gmelin***   **AMWI**
Uncommon summer resident in the Hudson Bay Lowland; occasional, rare to locally common elsewhere in summer, south to Lake Erie. Rare (to locally common) winter resident on the lower Great Lakes. Common (to locally abundant) migrant. Principal dates in province: mid March to mid November. Egg dates: 9 May to 19 August.

Canvasback, *Aythya valisineria* **(Wilson)***   **CANV**
Rare local summer resident in the south (Lake St. Clair and Luther Marsh), summer sightings north to Ottawa; possible breeder in the extreme west near Manitoba. Common (to locally abundant) winter resident in the lower Great Lakes. Abundant transient and migrant in the extreme south, uncommon elsewhere. Principal dates in province: early October to late April. Egg date: 24 May.

Redhead, *Aythya americana* **(Eyton)***   **REDH**
Common summer resident at Lake St. Clair; rare local elsewhere in the south (north to Sudbury); sight records in the north to Cochrane and Red Lake. Common (to locally abundant) winter resident on Lakes Ontario and Erie; occasional, rare elsewhere (north to Georgian Bay). Common (to locally abundant) migrant. Egg dates: 24 May to 4 July.

Ring-necked Duck, *Aythya collaris* **(Donovan)***   **RNDU**
Uncommon summer resident, north to Big Trout Lake and Attawapiskat, south to Luther Marsh and Kingston (almost entirely on the Canadian Shield). Rare winter resident in the south; vagrant in the north. Common (to locally abundant) migrant. Principal dates in province: mid March to mid November. Egg dates: 19 May to 28 July.

Tufted Duck, *Aythya fuligula* **(Linnaeus)**   **TUDU**
Occasional, rare straggler in the south, autumn to spring. First reported in 1956 (North, 1956); at least five reports in the past decade. Dates in province: November to April.

Greater Scaup, *Aythya marila* **(Linnaeus)***   **GRSC**
Uncommon summer resident on the Hudson Bay coast. Breeding may also occur occasionally on a few large lakes near Hudson Bay; summer records and suspected breeding on Lake Superior; occasional stragglers in summer on the lower Great Lakes (nonbreeding). Abundant winter resident on Lakes Ontario and Erie, rare north to Lake Superior. Abundant migrant. Egg dates: 2 to 14 July.
   A. m. nearctica.

Lesser Scaup, *Aythya affinis* **(Eyton)*** LESC
Uncommon, but local summer resident in the north; rare (to uncommon) in a few scattered locations in the south. Rare winter resident in the south. Common (to locally abundant) migrant. Principal dates in province: mid March to late November. Egg dates: 27 May to 11 July.

Common Eider, *Somateria mollissima* **(Linnaeus)*** COEI
Uncommon to locally common summer resident on the Hudson Bay coast. Permanent resident in the offshore waters of James and Hudson bays. Occasional, rare straggler in autumn, winter, and spring in the south (north to Sudbury). Egg dates: 26 June to 16 July.
    S. m. sedentaria breeds in the north; occasional, rare straggler in the south.
    S. m. dresseri occasional, rare straggler in the south.

King Eider, *Somateria spectabilis* **(Linnaeus)*** KIEI
Rare local breeding resident on the north coast. Probably rare to uncommon permanent resident in the offshore waters of Hudson and James bays. Vagrant in summer in the south. Rare straggler, autumn to spring in the south (north to Manitoulin Island and the Ottawa River). Egg date: 24 July.

Harlequin Duck, *Histrionicus histrionicus* **(Linnaeus)** HARD
Vagrant in summer. Rare winter resident in the south. Rare migrant on the Great Lakes and the north coast. Principal dates in province: late September to late April.

Oldsquaw, *Clangula hyemalis* **(Linnaeus)*** OLDS
Uncommon summer resident along the Hudson Bay coast, summer records from James Bay; nonbreeding stragglers on Lake Ontario and Lake Superior. Abundant winter resident on the southern Great Lakes. Abundant migrant. Egg dates: 10 to 25 July.

Black Scoter, *Melanitta nigra* **(Linnaeus)** BLSC
Abundant summer resident in waters off the north coast, possibly a widespread breeder but this has never been confirmed; vagrant on the Great Lakes (Thunder Bay) in summer (nonbreeding). Rare (to locally uncommon) winter resident on the lower Great Lakes. Common (to locally abundant) migrant. Principal dates in province: early April to mid November.
    M. n. americana.

Surf Scoter, *Melanitta perspicillata* **(Linnaeus)*** SUSC
Rare summer resident near the north coast; three breeding records within 100 km of the coast; common in offshore waters. Rare (to locally uncommon) winter resident on the lower Great Lakes. Common migrant. Principal dates in province: mid April to mid November. Egg dates: unknown.

White-winged Scoter, *Melanitta fusca* **(Linnaeus)*** WWSC
Rare local summer resident in the Hudson Bay Lowland; common in off-shore waters; occasional, rare straggler on the lower Great Lakes in summer (nonbreeding). Common winter resident on the lower Great Lakes. Common (to locally abundant) migrant. Principal dates in province: mid March to mid January. Egg date: 31 July.
 M. f. deglandi.

Common Goldeneye, *Bucephala clangula* **(Linnaeus)*** COGO
Common summer resident across the north (unlikely nesting north of about Big Trout Lake and Attawapiskat), and south to Manitoulin Island. Now virtually absent as a breeder elsewhere in the south, but summer sight records south to Lake Erie. Common (to abundant) winter resident in the south; occasional, uncommon in the west. Abundant migrant. Egg dates: 11 May to 21 July.
 B. c. americana.

Barrow's Goldeneye, *Bucephala islandica* **(Gmelin)** BAGO
Vagrant in summer in the south. Rare transient and winter resident. Principal dates in province: October to April.

Bufflehead, *Bucephala albeola* **(Linnaeus)*** BUFF
Rare, widely scattered summer resident in the north; occasional, rare, probably nonbreeding in summer in the south. Common winter resident on the lower Great Lakes. Common migrant. Egg dates: unknown.

Smew, *Mergellus albellus* **(Linnaeus)** SMEW
Vagrant; one photographic record: (ROM PR 109–110) Niagara River, 17 January to 30 March 1960 (Beardslee and Mitchell, 1965). Sight record: Normandale, Haldimand-Norfolk RM, 9 to 10 December 1973 (Goodwin, 1974a).

Hooded Merganser, *Lophodytes cucullatus* **(Linnaeus)*** HOME
Uncommon summer resident across the province, north rarely to Sandy Lake and Fort Albany; scattered and rare south of the Canadian Shield. Rare winter resident in the south (north to Georgian Bay and Ottawa). Common migrant. Principal dates in province: late March to late November. Egg dates: 8 April to 28 June.

Common Merganser, *Mergus merganser* **Linnaeus*** COME
Common summer resident across the province, except rare and occasional in the agricultural southern parts. Common (to locally abundant) winter resident on the lower Great Lakes; occasional, rare straggler in winter on Lake Superior. Common to locally abundant migrant. Egg dates: 18 April to 27 July.
 M. m. americanus.

Red-breasted Merganser, *Mergus serrator* **Linnaeus*** RBME
Uncommon summer resident across the north, especially in the Hudson Bay Lowland and on Lake Superior; uncommon on Georgian Bay and Lake Huron, rare south to Lake Erie. Uncommon winter resident in the south. Common to locally abundant migrant. Egg dates: 26 May to 6 August.
   M. s. serrator.

Ruddy Duck, *Oxyura jamaicensis* **(Gmelin)*** RUDU
Rare (to locally uncommon) summer resident in the agricultural southern parts of the province and near Sudbury; also possibly Thunder Bay (Denis, 1958) and Rainy River area; summer sight records, north to Cochrane. Occasional, rare (to locally common) winter resident in the south. Uncommon (to locally common) migrant. Principal dates in province: early April to mid November. Egg dates: 24 May to 20 August.
   O. j. rubida.

Black Vulture, *Coragyps atratus* **(Bechstein)** BLVU
Rare straggler in the south (north to the north shore of Lake Ontario), spring to autumn. Vagrant in winter. First reported in 1947 (Hope, 1949); at least eight reports in the past decade. Dates in province: (mid February) early May to mid October (early January).

Turkey Vulture, *Cathartes aura* **(Linnaeus)*** TUVU
Rare (to locally uncommon) summer resident in the west (Red Lake to Thunder Bay); uncommon in the south (north to Sault Ste. Marie and North Bay). Occasional, rare straggler in summer and autumn to the far north (to Hudson Bay). Occasional, rare straggler in winter in the south (north to Huntsville). Locally common (to abundant) migrant. Principal dates in province: late March to late October. Egg dates: 2 May to 15 July.
   C. a. septentrionalis breeds in the south.
   C. a. teter breeds in the west.

Osprey, *Pandion haliaetus* **(Linnaeus)*** OSPR
Rare (to locally uncommon) summer resident across the province, except largely absent in the agricultural areas off the Canadian Shield in the south. Occasional, rare straggler in the south in winter. Rare (to locally common) migrant. Principal dates in province: late April to late October. Egg dates: 9 May to 21 June.
   P. h. carolinensis.

American Swallow-tailed Kite, *Elanoides forficatus* **(Linnaeus)** ASTK
Occasional, rare straggler. Several records from the south dating back to 1860 (Fleming, 1907); one in the north (central Sudbury District). At least

six records since 1978, all but one in spring. Dates in province: early May to early September.
E. f. forficatus.

Mississippi Kite, *Ictinia mississippiensis* **(Wilson)**   **MIKI**
Occasional, rare straggler. Nearly a dozen records in the south (north to Toronto) dating back to 1951 (Baillie, 1952); at least six records since 1978, mainly in spring. Dates in province: early May to mid September.

Bald Eagle, *Haliaeetus leucocephalus* **(Linnaeus)***   **BAEA**
Rare to uncommon local summer resident in northwestern Ontario (west of about Big Trout Lake and Lake Nipigon); rare and very thinly scattered elsewhere, usually only near large lakes. Rare winter resident in the south; occasional in the west. Rare migrant. Principal dates in province: early March to mid October. Egg dates: 3 April to 28 June.
H. l. alascanus breeds.
H. l. leucocephalus rare vagrant.

Northern Harrier, *Circus cyaneus* **(Linnaeus)***   **NOHA**
Uncommon summer resident in most of the province; rare in the Boreal Forest Region and areas of intense agriculture. Rare winter resident in the south. Common migrant. Principal dates in province: early April to mid November. Egg dates: 26 April to 7 July.
C. c. hudsonius.

Sharp-shinned Hawk, *Accipiter striatus* **Vieillot***   **SSHA**
Rare to uncommon (but rarely seen) summer resident across the province, north to within about 100 km of Hudson Bay; becoming rare to absent in southern agricultural areas. Rare winter resident in the south; occasional, rare in winter in the north (Thunder Bay). Common migrant. Principal dates in province: early April to late October. Egg dates: 30 April to 30 June.
A. s. velox.

Cooper's Hawk, *Accipiter cooperii* **(Bonaparte)***   **COHA**
Rare summer resident, mainly on or near the Canadian Shield in southern Ontario; rare and widely scattered to absent south of the Shield. Probably rare and widely scattered in the north as far north as Kenora and Lake Abitibi; sight records to Sandy Lake and Attawapiskat. Rare winter resident in the south. Locally common migrant. Principal dates in province: early April to late October. Egg dates: 27 April to 8 July.

Northern Goshawk, *Accipiter gentilis* **(Linnaeus)***   **NOGO**
Rare summer resident in forested areas across the province, south to Lake Ontario, occasionally to Lake Erie. Permanent resident in the south, par-

tially migratory in much of the north. Rare to locally uncommon migrant. Egg dates: 16 April to 2 June.
  A. g. atricapillus.

Red-shouldered Hawk, *Buteo lineatus* **(Gmelin)***   RSHA
Rare (to locally uncommon) summer resident in the south, but absent from areas of extensive coniferous forests; now largely absent south of the latitude of Lake Ontario. Occasional, rare in the north to Wawa and Lake Abitibi. Occasional, rare straggler in winter in the south. Uncommon (to locally common) migrant. Principal dates in province: late March to late October. Egg dates: 1 April to 9 July.
  B. l. lineatus.

Broad-winged Hawk, *Buteo platypterus* **(Vieillot)***   BWHA
Uncommon summer resident across the province, north to Sandy Lake and Moosonee; rare and scattered south of the Canadian Shield. Abundant migrant. Principal dates in province: (late March) late April to mid October (late November). Egg dates: 22 April to 2 July.
  B. p. platypterus.

Swainson's Hawk, *Buteo swainsoni* **Bonaparte[*]**   SWHA
Rare transient, mainly in the south; also Moosonee (Duncan, 1986). Most often seen in autumn. One unconfirmed nest record: near Kenora (Peck and James, 1987). Dates in province: late April to late May; early September to late October. [Egg date: 9 June.]

Red-tailed Hawk, *Buteo jamaicensis* **(Gmelin)***   RTHA
Common summer resident south of Georgian Bay and Ottawa; becoming progressively rarer and more scattered throughout the Boreal Forest Region, almost to the Hudson Bay coast. Common winter resident in the south. Common (to locally abundant) migrant. Egg dates: 3 March to 15 July.
  B. j. borealis breeds over most of the province.
  B. j. kriderii rare (and occasional?) breeder in the extreme west (Kenora); wanders occasionally elsewhere.
  B. j. calurus occasional, rare wanderer throughout, autumn to spring.

Rough-legged Hawk, *Buteo lagopus* **(Pontoppidan)***   RLHA
Rare summer resident in northern tundra (breeding confirmed only in the Cape Henrietta Maria region); summer observations all along the James Bay coast. Vagrant (nonbreeding) south to Thunder Bay and Kingston in summer. Common winter resident in the south. Uncommon (to locally common) migrant. Principal dates in province: late August to late May. Egg dates: unknown.
  B. l. sanctijohannis.

Golden Eagle, *Aquila chrysaetos* **(Linnaeus)*** GOEA
Rare and very thinly scattered summer resident in the north (most records near the Hudson Bay coast); summer sight records on the Canadian Shield in the south, but only historical and unsubstantiated breeding records in the south. Rare winter resident in the south. Rare (to locally uncommon) migrant. Egg dates: unknown.
   A. c. canadensis.

Crested Caracara, *Polyborus plancus* **(Miller)** CRCA
Vagrant; one specimen: (now missing) Thunder Bay, 18 July 1892 (Atkinson, 1894). At the time and place, and with the described weather conditions, it seems unlikely to have been an escaped captive as suggested by Godfrey (1986). This record has been accepted by OBRC.
   [*P. p. audubonii.*]

American Kestrel, *Falco sparverius* **Linnaeus*** AMKE
Common summer resident south of Georgian Bay and Ottawa; becoming progressively rarer northward to Hudson Bay; absent from much of the Hudson Bay Lowland. Uncommon winter resident in the south. Common migrant. Principal dates in province: early March to late October. Egg dates: 11 April to 16 July.
   F. s. sparverius.

Merlin, *Falco columbarius* **Linnaeus*** MERL
Rare summer resident across the north (but may be absent from much of the densely wooded Boreal Forest Region), south to Manitoulin Island and Algonquin Park, very rarely to southern Georgian Bay. Rare winter resident in the south and in the west (Thunder Bay). Rare migrant. Egg dates: 16 May to 5 July.
   F. c. columbarius breeds.
   F. c. bendirei occasional, rare transient.
   [*F. c. richardsonii* possible sight record (Wormington, 1984b).]

Peregrine Falcon, *Falco peregrinus* **Tunstall*** PEFA
Former rare local summer resident, mainly in the south, and in the north to Lake Superior and the Timiskaming District; extirpated as a breeding bird by about 1964. Introductions of captive raised birds began in 1977. Nesting reported again in 1982, but probably still fewer than a dozen pairs within former range. Rare winter resident in the south. Rare to uncommon migrant. Principal dates in province: early April to late October. Egg dates: 23 April to 9 June.
   F. p. anatum breeds.
   F. p. tundrius transient.

Gyrfalcon, *Falco rusticolus* **Linnaeus**   **GYRF**
Rare transient during autumn and spring; probably regular on the north coast, occasional elsewhere. Rare winter resident. Dates in province: September to April (mid May).
    F. r. obsoletus.

[Prairie Falcon, *Falco mexicanus* **Schlegel**   **PRFA**
Four undocumented sight records from 1944 to 1962 (Speirs, 1958; Goodwin, 1962), at least some of which were escaped captives.]

Gray Partridge, *Perdix perdix* **(Linnaeus)***   **GRPA**
Rare to uncommon local permanent resident in southern agricultural areas (mainly in southeastern Ontario and in the Niagara Peninsula); formerly at Thunder Bay, at Sault Ste. Marie, and in Timiskaming District. Introduced. Egg dates: 7 May to 7 September.
    P. p. ? origins uncertain and probably mixed.

Ring-necked Pheasant, *Phasianus colchicus* **Linnaeus***   **RINP**
Common permanent resident in the extreme south (self-sustaining only in the Deciduous Forest Region); becoming rare and local north to Lake Simcoe and Ottawa, where populations are maintained through repeated releases. Has been released as far north as Thunder Bay. Introduced. Egg dates: 21 April to 28 August.
    P. c. ? origins uncertain and probably mixed.

Spruce Grouse, *Dendragapus canadensis* **(Linnaeus)***   **SPGR**
Uncommon (but rarely seen) permanent resident across the north (except Tundra areas); becoming rare in the south to Algonquin Park; occasional, rare to Simcoe and northern Hastings counties. Historical record (1908) from Peel RM (Baillie and Harrington, 1936). Egg dates: 9 May to 25 June.
    D. c. canadensis breeds across the north; intergrading extensively with
        canace south of Kenora and Cochrane.
    D. c. canace breeds in the south.

Willow Ptarmigan, *Lagopus lagopus* **(Linnaeus)***   **WIPT**
Rare to common (cyclical) permanent resident in northern Tundra areas. Occasional, rare in winter as far south as Moosonee; formerly irruptive south to Lake Nipissing; historical record (1897) from Durham RM (Fleming, 1907). Egg dates: 23 June to 17 July.
    L. l. albus.

Rock Ptarmigan, *Lagopus mutus* **(Montin)**   **ROPT**
Vagrant in summer; one report on the north coast (Weir, 1990b). Irruptive

in winter into far northern areas. Principal dates in province: December to late March.

L. m. rupestris.

Ruffed Grouse, *Bonasa umbellus* (Linnaeus)*   RUGR

Rare to common (cyclical) permanent resident throughout most of the province; becoming rare and local in the southern fringes of the Hudson Bay Lowland (absent in the northern Lowland); rare to absent in the extreme south (Lambton and Essex counties). Egg dates: 15 April to 17 July.

- *B. u. monticola* primarily restricted to the Deciduous Forest Region, intergrading extensively with *togata*.
- *B. u. togata* across the rest of southern Ontario, intergrading to the north with *obscura*.
- *B. u. obscura* across northern Ontario, intergrading with *togata* in the southern part of its range and with the prairie *umbelloides* form in the west.
- *B. u. incana* one specimen from extreme western Ontario (Rainy River District).

Greater Prairie-Chicken, *Tympanuchus cupido* (Linnaeus)*   GRPC

Extirpated, former resident in the extreme south (before 1900 in Essex and Kent counties). Invaded Kenora to Thunder Bay areas early in this century; gone by 1959. Invaded Sault Ste. Marie to Manitoulin Island area about 1925; gone by 1961 after extensive hybridization with Sharp-tailed Grouse (Lumsden, 1966, 1987b). Egg date: 20 June.

T. c. pinnatus.

Sharp-tailed Grouse, *Tympanuchus phasianellus* (Linnaeus)*   STGR

Rare to uncommon permanent resident across the north, south to Lake Nipigon and Lake Abitibi; also in the Lake of the Woods area, in southern Algoma District, and on Manitoulin Island. Isolated groups have been introduced in the south (to Prince Edward County). Egg dates: 9 June to 24 July.

- *T. p. phasianellus* breeds throughout the province, south to Lake Nipigon and Lake Abitibi, intergrading with *campestris* in the west.
- *T. p. campestris* in the Lake of the Woods area (formerly also near Thunder Bay), in southern Algoma District, and on Manitoulin Island; introduced elsewhere in the south.

Wild Turkey, *Meleagris gallopavo* Linnaeus*   WITU

Former common permanent resident in the south (north to Simcoe County and Durham RM), extirpated due to destruction of habitat (Clarke, 1943). Subsequent introductions within former range largely unsuccessful, except

where maintained by winter feeding, until 1984. Releases near Long Point and Napanee now expanding rapidly. Egg dates: 8 May to 25 July.
M. g. silvestris.

Northern Bobwhite, *Colinus virginianus* **(Linnaeus)***   **NOBO**
Uncommon permanent resident in the extreme south (north to Middlesex and Elgin counties, rare to southern Huron County and Hamilton); formerly to Muskoka and Kingston (Clarke, 1954). Released in several localities as far north as Ottawa. Egg dates: 21 May to 19 September.
C. v. virginianus.

Yellow Rail, *Coturnicops noveboracensis* **(Gmelin)***   **YERA**
Uncommon summer resident along the Hudson Bay and James Bay coasts; rare and local in the south and along the western edge of the province. Rare migrant. Dates in province: (late April) early May to late October. Egg dates: 12 to 30 June.
C. n. noveboracensis.

[Black Rail, *Laterallus jamaicensis* **(Gmelin)**   **BLRA**
Four undocumented sight records are mentioned in the literature between 1857 and 1959 (Cottle, 1859; Nash, 1894; Baillie, 1951; Speirs, 1959).]

King Rail, *Rallus elegans* **Audubon***   **KIRA**
Rare summer resident in the extreme south (near Lake Erie); occasional, rare north to the Bruce Peninsula and Ottawa. Occasional, rare straggler in winter in the south (northeast to Kingston). Rare migrant. Principal dates in province: early May to late October. Egg dates: 18 May to 17 July.
R. e. elegans.

Virginia Rail, *Rallus limicola* **Vieillot***   **VIRA**
Common summer resident in the south; occasional, rare north to Kenora and southern James Bay. Occasional, rare winter resident in the south (north to Kingston). Uncommon migrant. Principal dates in province: late April to early October. Egg dates: 4 May to 19 July.
R. l. limicola.

Sora, *Porzana carolina* **(Linnaeus)***   **SORA**
Common summer resident in the south; becoming rare and local on the Canadian Shield and in the north, but occasional, at least to Hudson Bay. Vagrant in winter (Quilliam, 1973). Uncommon migrant. Dates in province: (late March) mid April to late September (mid January). Egg dates: 4 May to 2 August.

Purple Gallinule, *Porphyrula martinica* **(Linnaeus)** PUGA
Occasional, rare straggler in the south (north to Ottawa), most often in autumn. Vagrant in the north (Moosonee). First reported in 1892 (Fleming, 1906); at least four reports in the past decade. Dates in province: early April to late October.

Common Moorhen, *Gallinula chloropus* **(Linnaeus)*** COMO
Uncommon (to locally common) summer resident in the south (north to Georgian Bay and Ottawa); occasional summer records as far north as Lake Superior. Occasional, rare straggler in winter in the extreme south. Common migrant. Principal dates in province: early May to mid October. Egg dates: 4 May to 21 July.
   G. c. cachinnans.

American Coot, *Fulica americana* **Gmelin*** AMCO
Rare (to locally common) summer resident, north to Georgian Bay and Ottawa, rare north to Red Lake and Kapuskasing; sight records north to Lake River and Fort Severn. Rare winter resident in the south (north to Manitoulin Island). Common (to locally abundant) migrant, mainly in the south; vagrant to James Bay. Principal dates in province: late March to late November. Egg dates: 4 May to 16 July.
   F. a. americana.

Sandhill Crane, *Grus canadensis* **(Linnaeus)*** SACR
Rare to locally uncommon summer resident in the Hudson Bay Lowland, and south to the Cochrane area; also, on eastern Lake Superior to Sudbury, on Manitoulin Island, at the top of the Bruce Peninsula, in the Ottawa area, and in western Rainy River District; formerly in Essex and Kent counties. Populations currently increasing. Occasional, rare winter resident in the south. Rare migrant in the south; rare to common migrant in the west. Principal dates in province: early April to early October. Egg dates: 5 May to 4 June.
   G. c. tabida probably the form breeding in the Rainy River area, in the Sault Ste. Marie to Sudbury area, on Manitoulin Island, on the Bruce Peninsula, and in the Ottawa area; formerly in Essex and Kent counties.
   G. c. canadensis probably a rare migrant (Lumsden, 1971).
   G. c. rowani possibly the form breeding in the Hudson Bay Lowland, and south to about Pickle Lake and Cochrane.

Whooping Crane, *Grus americana* **(Linnaeus)** WHCR
Former rare (?) migrant in the province; one specimen: (Queen's University) Yarker, Lennox and Addington County, September 1871 (McIlwraith, 1894). Archaeological material (ROM 94282) from Fort Albany. Sight records: Toronto, prior to 1887 (Chamberlain, 1887); Parry Sound District, l895 (Fleming, 1901); Long Point, June 1898 (Macoun, 1898).

Black-bellied Plover, *Pluvialis squatarola* **(Linnaeus)** BBPL
Common (to locally abundant) spring and autumn, transient throughout the province. Dates in province: (early April) May (late June); (July) August to late October (mid December).

Lesser Golden-Plover, *Pluvialis dominica* **(Müller)*** LEGP
Rare to uncommon summer resident in the Cape Henrietta Maria region; occasional elsewhere along the Hudson Bay coast. Unexpected elsewhere in summer, but a few records of nonbreeding birds south to Point Pelee. Rare to locally common spring, and uncommon to locally abundant autumn, migrant and transient. Dates in province: (mid March) mid April to late September (mid December). Egg dates: 23 June to 8 July.
 P. d. dominica.

Mongolian Plover, *Charadrius mongolus* **Pallas** MONP
Vagrant; one photographic record: (ROM PR 1569–1570) Presqu'ile Provincial Park, 4 May 1984 (McRae, 1985).

Snowy Plover, *Charadrius alexandrinus* **Linnaeus** SNPL
Vagrant; one photographic record: Long Point, 4 to 9 May 1987 (Collier and Curson, 1988); also a report from Long Point, 9 May 1990 (Weir, 1990a). A specimen from Toronto, May 1880, reported by Fleming (1906), was destroyed and no description exists. Another specimen (now missing) mentioned by Ames (1897) was from an unreliable collector.

Wilson's Plover, *Charadrius wilsonia* **Ord** WIPL
Vagrant; one documented record: Hamilton, 26 to 31 May 1990 (Weir 1990a); one record for which documentation has been lost (Goodwin, 1966); and a possible occurrence mentioned by Stirrett (1973a).

Semipalmated Plover, *Charadrius semipalmatus* **Bonaparte*** SEPL
Uncommon summer resident along the Hudson Bay coast and adjacent interior; becoming rare southward along the James Bay coast to the Moosonee area. Common migrant. Dates in province: (mid April) early May to early October (early December). Egg dates: 5 June to 30 July.

Piping Plover, *Charadrius melodus* **Ord*** PIPL
Now rare (and occasional?) summer resident on Lake of the Woods. Apparently gone from the south, where it was formerly a rare to locally common nester on the shores of the Great Lakes. Rare migrant. Dates in province: (early April) early May to mid September (mid November). Egg dates: 8 May to 29 July.

Killdeer, *Charadrius vociferus* **Linnaeus***   **KILL**
Common summer resident in the south; becoming rare and local in the north to the Hudson Bay coast. Rare in winter in the south (north to Ottawa). Common migrant. Principal dates in province: mid March to late October. Egg dates: 1 April to 21 August.
   C. v. vociferus.

American Oystercatcher, *Haematopus palliatus* **Temminck**   **AMOY**
Vagrant; one specimen: (ROM 90937) Niagara RM, 21 July 1960 (Lunn, 1961). One sight record: Hamilton area, 2 November 1985.
   H. p. palliatus.

Black-necked Stilt, *Himantopus mexicanus* **(Müller)**   **BNST**
Vagrant; one photographic record: Stoney Point, Essex County, 28 May to 16 June 1989 (Weir, 1989d; Wormington and Curry, 1990). Four sight records with accepted descriptions: Timmins, 1 September l955 (Baillie, l955); Smithville, Niagara RM, 14 October 1979; Sable Island, Lake of the Woods, 7 June 1981; North Bay, 12 to 23 June 1989.

American Avocet, *Recurvirostra americana* **Gmelin***   **AMAV**
Rare summer resident; one breeding record at Lake of the Woods, July 1980. Rare (to locally uncommon) transient, mainly in the south. Dates in province: (mid April) early May to mid October (mid December). Egg dates: unknown.

Greater Yellowlegs, *Tringa melanoleuca* **(Gmelin)***   **GRYE**
Common summer resident in the Hudson Bay Lowland, rare south to Red Lake, Jellicoe, and Cochrane. Common (to locally abundant) migrant. Dates in province: (mid March) mid April to late October (late December). Egg dates: 12 June to 1 July.

Lesser Yellowlegs, *Tringa flavipes* **(Gmelin)***   **LEYE**
Common summer resident in the northern Hudson Bay Lowland, rare south to Sandy Lake and Moosonee. Common (to locally abundant) migrant. Dates in province: (early March) mid April to mid October (mid December). Egg dates: 4 to 12 June.

Spotted Redshank, *Tringa erythropus* **(Pallas)**   **SPRD**
Vagrant; two photographic records: (ROM PR 1342) Peterborough County, 8 May 1981 (James, l983); Casselman, Prescott and Russell County, 19 to 24 July 1990 (Weir, 1990b). One sight record: Niagara RM, 25 July 1976 (Axtell, Benham, and Black, 1977).

Solitary Sandpiper, *Tringa solitaria* **Wilson*** SOSA
Uncommon summer resident throughout the north. Uncommon migrant.
Dates in province: (mid March) early May to late September (early December). Egg date: 28 June.
   T. s. solitaria.

Willet, *Catoptrophorus semipalmatus* **(Gmelin)** WILL
Occasional reports of birds in summer are likely nonbreeding transients. Rare (to locally uncommon) spring and autumn transient, most often seen in the south; more numerous in spring. Dates in province: (mid April) early May to early June; early July to late September (late October).
   C. s. inornatus.
   [*C. s. semipalmatus* single sight record at Long Point, autumn 1985 (Weir, 1986a).]

Wandering Tattler, *Heteroscelus incanus* **(Gmelin)** WATA
Vagrant; one photographic record: (ROM PR 829–830) Niagara RM, 12 June 1977 (James, 1983). Two sight records: Niagara RM, 1 August 1948, and 12 to 15 July 1960 (Beardslee and Mitchell, 1965).

Spotted Sandpiper, *Actitis macularia* **(Linnaeus)*** SPSA
Common summer resident throughout the province. Vagrant in winter (Weir, 1982, 1987a). Common migrant. Dates in province: (late March) early May to late September (mid January). Egg dates: 14 May to 23 July.

Upland Sandpiper, *Bartramia longicauda* **(Bechstein)*** UPSA
Uncommon summer resident in the south (but rare on the Canadian Shield); rare in the west (Thunder Bay and Rainy River District). Uncommon migrant. Dates in province: (early April) late April to mid September (late October). Egg dates: 12 May to 9 July.

Eskimo Curlew, *Numenius borealis* **(Forster)** ESCU
Now nearly extinct and unlikely to be seen. Former regular autumn (and probably spring) transient, more numerous in the north; probably common to abundant (Gollop, Barry, and Iverson, 1986). Dates in province: April to May; August to October.

Whimbrel, *Numenius phaeopus* **(Linnaeus)*** WHIM
Rare local summer resident along the coasts of Hudson Bay and northern James Bay. Uncommon (to locally abundant) migrant and transient. Dates in province: (early May) mid May to mid September (late November). Egg dates: 25 June to 17 July.
   N. p. hudsonicus.

Slender-billed Curlew, *Numenius tenuirostris* **Vieillot** SBCU
Vagrant; one specimen: (BSNS 2092) Niagara RM, October 1925 (Sheppard, 1960; Beardslee and Mitchell, 1965; James 1983).

Long-billed Curlew, *Numenius americanus* **Bechstein** LBCU
Vagrant; one photographic record: (ROM PR 195–197) Ajax, Durham RM, 17 October 1959 (Speirs, 1960a), has been accepted by OBRC (Wormington, 1987). A sight record at Kettle Point, 24 July 1983 (Weir, 1983b), remains controversial. Although mentioned as an "irregular visitor" by McIlwraith (1886), there is no specimen evidence of its occurrence at that time.

Hudsonian Godwit, *Limosa haemastica* **(Linnaeus)*** HUGO
Rare (to locally uncommon) summer resident along the coasts of Hudson Bay and the northern half of James Bay. Common (to abundant) migrant and transient in the north, rare (to locally common) in the south and west. Dates in province: (late April) mid May to mid October (late November). Egg dates: unknown.

Marbled Godwit, *Limosa fedoa* **(Linnaeus)*** MAGO
Rare (to locally uncommon) summer resident along the James Bay coast; possibly also occasional in western Rainy River District, but never verified. Rare migrant. Dates in province: (mid April) mid May to mid September (late September). Egg date: 1 July.

Ruddy Turnstone, *Arenaria interpres* **(Linnaeus)** RUTU
Uncommon (to locally abundant) transient throughout the province. Dates in province: (mid April) mid May to late June; mid July to late September (late December).
 A. i. morinella (only ?).

Red Knot, *Calidris canutus* **(Linnaeus)** REKN
Uncommon (to locally abundant) spring and autumn transient, most numerous on the north coast. Dates in province: (early May) mid to late May (June); (early July) September (mid November).
 C. c. rufa.

Sanderling, *Calidris alba* **(Pallas)** SAND
Common spring and autumn transient throughout the province. Dates in province: (mid April) May (late June); (early July) September to October (late December). Although Swainson and Richardson (1831) list it as a summer resident on the north coast, they likely observed only transients.

Semipalmated Sandpiper, *Calidris pusilla* **(Linnaeus)*** SESA
Common summer resident in the Cape Henrietta Maria region, rare to uncommon along the rest of the coast of Hudson Bay. Common (to abun-

dant) migrant and transient. Dates in province: (late April) mid May to early October (late November). Egg dates: 24 to 30 June.

Western Sandpiper, *Calidris mauri* **(Cabanis)** WESA
Rare (to locally common) spring and autumn transient in the south. Dates in province: early May to early June; (mid July) August to September (early December).

[Rufous-necked Stint, *Calidris ruficollis* **(Pallas)** RNST
Three sight records: Toronto, 1976; Point Pelee, 1983; Erieau, 1986: all rejected by OBRC (James, 1982, 1984b; Wormington, 1987.]

Little Stint, *Calidris minuta* **(Leisler)** LIST
Vagrant; one specimen: (NMC 68651) North Point, James Bay, 10 July 1979 (Morrison, 1980).

Least Sandpiper, *Calidris minutilla* **(Vieillot)*** LESA
Common summer resident along the Hudson Bay coast; becoming rare in the northern Hudson Bay Lowland, and south along the James Bay coast to Moosonee. Common migrant. Dates in province: (early April) early May to late September (late December). Egg dates: 22 June to 21 July.

White-rumped Sandpiper, *Calidris fuscicollis* **(Vieillot)** WRSA
Common spring and autumn transient on the north coast; rare to uncommon elsewhere in the north; locally common in the south in autumn, rare in spring. Dates in province: (early May) mid May to early June (late June); (early July) August to October (early December).

Baird's Sandpiper, *Calidris bairdii* **(Coues)** BASA
Rare spring and uncommon autumn transient. Dates in province: (mid April) mid May to early June; (mid July) August to mid September (late December).

Pectoral Sandpiper, *Calidris melanotos* **(Vieillot)*** PESA
Rare summer resident in the Cape Henrietta Maria region. Probably occasional elsewhere along the Hudson Bay coast, and more numerous on occasion in the Cape region. Vagrant in winter (Weir, 1985a). Common (to locally abundant) migrant and transient. Dates in province: (mid March) early April to late October (early January). Egg dates: unknown.

Sharp-tailed Sandpiper, *Calidris acuminata* **(Horsfield)** SHSA
Vagrant; two accepted sight records: Dundas, Hamilton-Wentworth RM, 19 November to 5 December 1975 (Wormington, 1986); Essex County, 20 August 1988 (Weir, 1989a; Wormington and Curry, 1990).

Purple Sandpiper, *Calidris maritima* (Brünnich)   **PUSA**
Occasional, rare in summer on the north coast and in the south (nonbreeding). Occasional, rare in winter in the south (north to Ottawa). Usually rare (to locally uncommon) autumn transient on Lake Ontario and the north coast; occasional, rare spring transient. Principal dates in province: late October to late December.

Dunlin, *Calidris alpina* (Linnaeus)*   **DUNL**
Common summer resident in the Cape Henrietta Maria and extreme northwestern coastal areas, less numerous along the rest of the Hudson Bay coast; unlikely elsewhere, but records of nonbreeding birds south to Lake Erie. Vagrant in winter (Goodwin, 1980a). Common to abundant migrant and transient. Dates in province: (late March) late April to early November (early January). Egg dates: 23 June to 20 July.
    C. a. pacifica.

Curlew Sandpiper, *Calidris ferruginea* (Pontoppidan)   **CUSA**
Occasional, rare straggler in the south, usually in spring and autumn. First reported in 1886 (Fleming, 1906); at least 12 reports in the past decade. Dates in province: mid May to late October.

Stilt Sandpiper, *Calidris himantopus* (Bonaparte)*   **STSA**
Common summer resident in the Cape Henrietta Maria region, rare to uncommon elsewhere along the Hudson Bay coast. Rare (to locally uncommon) migrant and transient in spring, uncommon (to locally common) in autumn. Dates in province: early May to mid September (early November). Egg dates: unknown.

Buff-breasted Sandpiper, *Tryngites subruficollis* (Vieillot)   **BBSA**
Rare (to locally uncommon) spring and autumn transient in the south and west, uncommon on the north coast. Dates in province: mid May to early June; late July to mid September (late October).

Ruff, *Philomachus pugnax* (Linnaeus)   **RUFF**
Rare transient, spring to autumn. Becoming increasingly numerous in recent years. Dates in province: (mid April) early May to mid June; early July to late September (early November).

Short-billed Dowitcher, *Limnodromus griseus* (Gmelin)[*]   **SBDO**
Rare summer resident along the Hudson Bay coast, possibly in the northern Hudson Bay Lowland and along the James Bay coast (Tuck, 1968; Cadman, Eagles, and Helleiner, 1987). Common transient and migrant. Dates in

province: (late April) mid May to mid October (late November). Egg dates: unknown.
  L. g. hendersoni transient and probable breeder.
  L. g. griseus transient and probable breeder.

Long-billed Dowitcher, *Limnodromus scolopaceus* **(Say)**   **LBDO**
Rare (to locally uncommon) transient, more usual in the south and in the autumn. Dates in province: (early April) early to mid May; late July to late October (early December).

Common Snipe, *Gallinago gallinago* **(Linnaeus)***   **COSN**
Common summer resident throughout most of the province (uncommon in areas of boreal forest, and rare to absent in the extreme south). Rare winter resident in the south (also at Atikokan; Peruniak, 1971). Common migrant. Principal dates in province: early April to mid November. Egg dates: 24 April to 19 July.
  G. g. delicata.

American Woodcock, *Scolopax minor* **Gmelin***   **AMWO**
Common summer resident in the south; becoming rare to locally uncommon in the north to Kenora, Lake Nipigon, and Cochrane; sight records north to Fort Albany. Occasional, rare in winter in the south. Uncommon migrant; wanders north to Moosonee. Principal dates in province: late March to early November. Egg dates: 2 April to 11 July.

Wilson's Phalarope, *Phalaropus tricolor* **(Vieillot)***   **WIPH**
Uncommon local summer resident in the south, along the southern James Bay coast, and in the Rainy River area; sight records near the western border, north to Sandy Lake. Populations increasing in recent years. Rare (to locally common) migrant. Dates in province: (late March) early May to early September (early December). Egg dates: 18 May to 4 July.

Red-necked Phalarope, *Phalaropus lobatus* **(Linnaeus)***   **RNPH**
Common summer resident in the Cape Henrietta Maria region, rare elsewhere along the Hudson Bay coast; probably occasional, and rare south almost to Moosonee along the James Bay coast. Rare in spring to uncommon in autumn, migrant and transient. Dates in province: (mid April) late May to late September (late November). Egg dates: 22 June to 4 July.

Red Phalarope, *Phalaropus fulicaria* **(Linnaeus)**   **REPH**
Vagrant in winter (Goodwin, 1980a). Rare transient in the south in autumn; vagrant in spring. Dates in province: May; (late August) mid September to mid November (early January).

Pomarine Jaeger, *Stercorarius pomarinus* **(Temminck)** POJA
Rare transient along the Hudson Bay coast, and in the southern Great Lakes and Ottawa River areas, mainly during autumn, occasionally in spring. Dates in province: May and late August to early November (early January).

Parasitic Jaeger, *Stercorarius parasiticus* **(Linnaeus)*** PAJA
Rare summer resident along the Hudson Bay coast; vagrant in the south (nonbreeding). Rare in winter in the south. Rare (to locally uncommon) migrant and transient in the south, mainly in autumn, occasionally in spring. Egg dates: 25 to 26 June.

Long-tailed Jaeger, *Stercorarius longicaudus* **Vieillot** LTJA
Occasional, rare in summer along the Hudson Bay coast; vagrant in the south (nonbreeding). Occasional, rare autumn transient on the Great Lakes. Dates in province: May to November.

[Great Skua, *Catharacta skua* **Brünnich** GRSK
Two reports of specimens (both now missing): Niagara RM, spring 1866 (Bergtold, 1889), and 3 December 1915 (Reinecke, 1916a); and two other sightings. None accepted by OBRC.]

Laughing Gull, *Larus atricilla* **Linnaeus** LAGU
Rare straggler on the lower Great Lakes; occasional, rare straggler in the north to the Hudson Bay coast, mainly in spring and autumn. Dates in province: (early March) early May to early October (early January).

Franklin's Gull, *Larus pipixcan* **Wagler** FRGU
Uncommon to abundant in nonbreeding flocks, in summer at Lake of the Woods. Occasional, rare in winter in the south. Rare (to common) transient in the extreme west; occasional, rare elsewhere in the north (to Moosonee); locally rare (to uncommon) transient in the south. Principal dates in province: early May to late October.

Little Gull, *Larus minutus* **Pallas*** LIGU
Rare local summer resident on the southern Great Lakes and in the Hudson Bay Lowland. Rare (to locally uncommon) winter resident in the south. Rare (to locally common) migrant. Principal dates in province: late April to late November. Egg dates: 27 May to 7 July.

Common Black-headed Gull, *Larus ridibundus* **Linnaeus** CBHG
Rare straggler in the south (Lakes Erie and Ontario) any month of the year, but mainly May to October.
  L. r. ridibundus.

Bonaparte's Gull, *Larus philadelphia* (Ord)* BOGU
Locally uncommon summer resident in the north, south to Lake Nipigon and Cochrane; nonbreeding birds occasional, rare south to Lake Erie. Common (to abundant) in early winter on the lower Great Lakes, becoming rare to absent by mid February. Common to abundant migrant. Principal dates in province: late March to late January. Egg dates: 29 May to 19 July.

Mew Gull, *Larus canus* Linnaeus MEGU
Occasional, rare straggler to southern Ontario, mainly in autumn. First reported in 1967 (Mitchell and Andrle, 1970); at least 10 reports in the past decade. Dates in province: mid February to late June (late July); late August to late November.
 L. c. brachyrhynchus.

Ring-billed Gull, *Larus delawarensis* Ord* RBGU
Abundant summer resident north to Lake of the Woods and Lake Superior; isolated small colonies north to Sachigo Lake and Attawapiskat; sight records from the north coast. Common winter resident in the south. Common to abundant migrant. Egg dates: 14 April to 30 July.

California Gull, *Larus californicus* Lawrence[*] CAGU
Unexpected summer resident; an unpaired female incubating eggs was present at Toronto in 1981 and 1982 (Blokpoel, 1987), providing the first accepted record of the species in Ontario. Occasional, rare straggler in other seasons, mainly in the south; also Moosonee. Nearly a dozen reports since 1980. Dates in province: (mid April) May to June; early September to late December (mid January). Egg dates: 14 May to 2 June.

Herring Gull, *Larus argentatus* Pontoppidan* HERG
Common summer resident throughout the province. Common (to locally abundant) winter resident on the southern Great Lakes; rare on Lake Superior. Common (to locally abundant) migrant. Egg dates: 14 April to 27 July.
 L. a. smithsonianus.

Iceland Gull, *Larus glaucoides* Meyer ICGU
Vagrant in summer (south or north). Rare (to locally common) in winter in the south, and the west (Thunder Bay). Rare (to uncommon) transient in the south and on the north coast (occasional elsewhere in the north). Principal dates in province: mid October to mid April.
 L. g. glaucoides.
 L. g. kumlieni.
 L. g. thayeri.

Lesser Black-backed Gull, *Larus fuscus* **Linnaeus** LBBG
Rare summer resident (nonbreeding). Rare (to uncommon) transient and winter resident in the south; occasional in the north (Wawa and James Bay). First reported in 1971, increasing steadily in abundance at present. Principal dates in province: mid October to May.
  L. f. graellsii usual.
  [*L. f. fuscus* two sight records (e.g., Weir, 1987b).]

Glaucous Gull, *Larus hyperboreus* **Gunnerus** GLGU
Occasional, rare straggler in summer, mainly on the north coast. Rare (to locally common) winter resident in the Great Lakes region. Rare to uncommon transient. Principal dates in province: early November to late April.
  L. h. hyperboreus (= *leuceretes* of Banks, 1986).

Great Black-backed Gull, *Larus marinus* **Linnaeus*** GBBG
Rare local summer resident on the lower Great Lakes; occasional, rare straggler in summer on the north coast, also on Lake Superior. Increasing slowly in abundance. Uncommon (to common) winter resident on the lower Great Lakes. Uncommon (to locally abundant) migrant and transient in the south and on the north coast. Egg dates: 7 June to 19 July.

Black-legged Kittiwake, *Rissa tridactyla* **(Linnaeus)** BLKI
Vagrant in summer in the south. Occasional, rare straggler in winter in the south. Rare to locally uncommon transient in spring and autumn (more often in autumn) in the south. Principal dates in province: late April to late May; mid September to late December.
  R. t. tridactyla.

Ross' Gull, *Rhodostethia rosea* **(MacGillivray)** ROGU
Vagrant; one photographic record: (ROM PR 1418–1426, 1493–1496) Moosonee, 14 to 24 May 1983 (Abraham, 1984). No other accepted records.

Sabine's Gull, *Xema sabini* **(Sabine)** SAGU
Vagrant in summer on the north coast. Rare (to locally uncommon) transient in the south in autumn; vagrant in spring. Dates in province: (late May); (mid July) early September to early November (mid December).
  X. s. sabini.

Ivory Gull, *Pagophila eburnea* **(Phipps)** IVGU
Occasional, rare straggler in autumn and winter throughout the province, probably more often in the far north, although most sightings are in the south. Reported since the late 1800s (McIlwraith, 1894); at least four reports in the past decade. Dates in province: (mid November) early December to late February (mid March).

Caspian Tern, *Sterna caspia* **Pallas***   CATE
Locally common summer resident on larger lakes in the south; rare on James Bay; summer sightings on Hudson Bay and Lake of the Woods. Uncommon migrant. Dates in province: (early April) late April to late September (early December). Egg dates: 13 May to 19 July.

Royal Tern, *Sterna maxima* **Boddaert**   ROYT
Vagrant; one sight record: Kingsville, 22 August 1974 (James, 1984b), accepted by OBRC.
[*S. m. maxima.*]

Sandwich Tern, *Sterna sandvicensis* **Latham**   SATE
Vagrant; one specimen: (ROM 91.10.1.58) Bruce County, autumn 1881 (McIlwraith, 1886). Three sight records in 1988: Hamilton, 24 April; Long Point, 17 May (Weir, 1988b); Presqu'ile Provincial Park, 14 to 25 June (Weir, 1988c).
S. s. acuflavida.

[Roseate Tern, *Sterna dougallii* **Montagu**   ROST
There are no specimens from Ontario; no descriptions of several records from 1886 to the present, either published (Sheppard, 1960; Stirrett, 1973a; Goodwin, 1972a; Weir, 1985b) or unpublished, have been accepted by OBRC.]

Common Tern, *Sterna hirundo* **Linnaeus***   COTE
Locally uncommon to common summer resident across the province, north almost to the Hudson Bay coast. Common (to locally abundant) migrant. Dates in province: (late March) early May to late September (mid December). Egg dates: 8 May to 17 August.
S. h. hirundo.

Arctic Tern, *Sterna paradisaea* **Pontoppidan***   ARTE
Uncommon summer resident within 100 km of the Hudson Bay coast and in the area near Akimiski Island; occasional summer records of nonbreeding birds, south to Lake Erie. Uncommon spring migrant along the eastern edge of the province (Godfrey, 1973); occasional, rare elsewhere. Dates in province: (mid May) late May to mid August. Egg dates: 16 June to 16 July.

Forster's Tern, *Sterna forsteri* **Nuttall***   FOTE
Locally uncommon to common summer resident, mainly on Lake St. Clair and Lake Erie, also on southern Lake Huron; summer sightings from western Ontario (Red Lake and Lake of the Woods). Rare (to locally common) migrant on the lower Great Lakes. Dates in province: (late March) mid April to late September (early December). Egg dates: 22 May to 26 June.

Least Tern, *Sterna antillarum* **(Lesson)** LETE
Vagrant; one sight record: Niagara RM, 26 June 1958 (Beardslee and Mitchell, 1965).

Sooty Tern, *Sterna fuscata* **Linnaeus** SOTE
Vagrant; one sight record: Leeds and Grenville County, 14 August l955 (Baillie, 1955), accepted by OBRC.

Black Tern, *Chlidonias niger* **(Linnaeus)*** BLTE
Locally uncommon to common summer resident in the south, rare to uncommon in the north as far north as Sandy Lake and Fort Albany. Common migrant. Dates in province: (early April) early May to late August (late November). Egg dates: 17 May to 24 July.
 C. n. surinamensis.

Black Skimmer, *Rynchops niger* **Linnaeus** BLSK
Vagrant; two photographic records: (ROM PR 1068–1070) Whitby, 1 November 1977 (Bain, 1978); (ROM PR 1286) Erieau, Kent County, 11 to 15 September 1981. Two sight records: Point Pelee, 6 July 1978; Lake of the Woods, late August 1982.
 [*R. n. niger.*]

Dovekie, *Alle alle* **(Linnaeus)** DOVE
Occasional, rare straggler in autumn and winter; seven records between 1924 and 1988. Dates in province: late October to early February.
 A. a. alle.

Thick-billed Murre, *Uria lomvia* **(Linnaeus)** TBMU
Occasional, rare to common straggler in the south, mainly in autumn and early winter. Three to four decades may pass between large flights. Last large flight 1950; last report 1953. Principal dates in province: late November to late December.
 U. l. lomvia.

Razorbill, *Alca torda* **Linnaeus** RAZO
Occasional, rare straggler to Lake Ontario (also in Lanark and Renfrew counties), almost any time of year, but mainly late October to late December. First reported in 1889 (Thompson, 1890); at least five reports in the past decade.
 A. t. torda.

Black Guillemot, *Cepphus grylle* **(Linnaeus)*** BLGU
Occasional, uncommon to common in summer, in offshore waters along the north coast. Only one breeding record, west of Cape Henrietta Maria (Lumsden, 1959). Vagrant in the south (Lake Ontario) in winter (also one

June record); none since 1954. Nonmigratory, but moves offshore to open waters in winter. Egg dates: unknown.
   C. g. ultimus.

Ancient Murrelet, *Synthliboramphus antiquus* **(Gmelin)**   **ANMU**
Vagrant; two specimens: (ROM 27.5.2.319) Toronto, 18 November 1901 (Fleming, 1906); (ROM 39908) Fort Erie, 15 November 1908 (Fleming, 1912).

Atlantic Puffin, *Fratercula arctica* **(Linnaeus)**   **ATPU**
Vagrant; one specimen (now missing): Ottawa, October 1881 (Lloyd, 1923). One photographic record: Ottawa, 15 December 1985 (DiLabio and Bouvier, 1986; Weir, 1986b).

Rock Dove, *Columba livia* **Gmelin***   **RODO**
Common permanent resident in southern agricultural areas; becoming local and uncommon in settled parts of the province, north to Kenora and Cochrane. Egg dates: 1 January to 31 December. Introduced many times; no subspecies recognized because of mixed origins.

Band-tailed Pigeon, *Columba fasciata* **Say**   **BTPI**
Occasional, rare straggler, mainly in the south. First reported in 1930 (Bell, 1941); at least six subsequent records, two in the north (at Dorion and at Wabigoon).
   C. f. monilis.

White-winged Dove, *Zenaida asiatica* **(Linnaeus)**   **WWDO**
Vagrant; one specimen: (ROM 67776) Fort Albany, 17 June 1942 (Shortt and Hope, 1943). One photographic record: Belleville, 14 to 19 December 1975. One sight record: Thunder Bay, 26 to 27 April 1986.
   Z. a. mearnsi.

Mourning Dove, *Zenaida macroura* **(Linnaeus)***   **MODO**
Common summer resident in southern agricultural areas; becoming rare, north to Sioux Lookout and Cochrane; wandering north to the Hudson Bay coast. Locally common winter resident in the agricultural south; occasional, rare north to Thunder Bay, Sault Ste. Marie, and Englehart. Common migrant. Egg dates: 19 March to 28 September.
   Z. m. carolinensis breeds across the province.
   Z. m. marginella breeds across the province, possibly more usual in the west.

Passenger Pigeon, *Ectopistes migratorius* **(Linnaeus)***   **PAPI**
Extinct: formerly an abundant summer resident, north to at least Kenora and Moosonee, possibly farther; at least occasional, and rare to uncommon

in winter; and an abundant migrant (Mitchell, 1935). Last specimen taken in September 1891 (Reinecke, 1916b), and last sight record was in May 1902 (Fleming, 1903). Principal dates in province: April to October. Egg dates: probably May to June.

Common Ground-Dove, *Columbina passerina* **(Linnaeus)** COGD
Vagrant; one specimen: (ROM 103396) Red Rock, Thunder Bay District, 29 October 1968 (Dick and James, 1969).
   C. p. passerina.

[Monk Parakeet, *Myiopsitta monachus* **(Boddaert)** MOPA
Single birds at Point Pelee National Park, 13 to 15 May and 21 June 1984, may have been vagrant feral birds or escaped pets (Wormington, 1984a). Other sightings are probably of escaped captives.]

[Carolina Parakeet, *Conuropsis carolinensis* **(Linnaeus)** CARP
A brief account suggesting a possible occurrence is given by Saunders and Dale (1933). Bones were found in 1984 at an archaeological site near London; however, evidence suggests that the bones were from a skin of decorative or ceremonial significance and thus may have been carried into Ontario (Prevec, 1985).]

Black-billed Cuckoo, *Coccyzus erythropthalmus* **(Wilson)*** BBCU
Uncommon to locally common summer resident in southern agricultural areas; becoming rare on the Canadian Shield, north to Sioux Lookout, Lake Nipigon, and Cochrane; occasional north to Moosonee. Common migrant. Dates in province: (late April) mid May to mid September (early November). Egg dates: 20 May to 8 September.

Yellow-billed Cuckoo, *Coccyzus americanus* **(Linnaeus)*** YBCU
Rare to uncommon summer resident in southern agricultural areas; rare north to Sudbury; occasional north to Atikokan, Thunder Bay, and Wawa. Uncommon migrant (vagrant north to Moosonee). Dates in province: (early May) mid May to mid September (mid November). Egg dates: 23 May to 7 August.
   C. a. americanus.

Groove-billed Ani, *Crotophaga sulcirostris* **Swainson** GBAN
Vagrant; two specimens: (ROM 132031) Sundridge, Parry Sound District, 27 October 1978; (ROM 151473) Rosslyn, Thunder Bay District, 1 November 1983. One photographic record: Sarnia, 9 to 13 October 1988 (Weir, 1989a). Two sight records: Stromness, Haldimand-Norfolk RM, 12 October 1969 (Goodwin, 1970); Red Rock, Thunder Bay District, 18 to 20 October 1963 (Wormington, 1986).
   C. s. sulcirostris.

Barn Owl, *Tyto alba* **(Scopoli)***   BROW
Rare local permanent resident in the extreme south (north to London and Hamilton; also occasional in the Kingston to Brockville area); vagrant north to Wawa and Ottawa. Has been found at Thunder Bay, but on the front of a train. Partially migratory. Egg dates: 3 March to 15 September.
   T. a. pratincola.

Eastern Screech-Owl, *Otus asio* **(Linnaeus)***   EASO
Uncommon (to locally common) permanent resident in southern agricultural areas; rare north to Sault Ste. Marie, Manitoulin Island, and Ottawa; sight records from Kenora and Sudbury. Egg dates: 20 March to 7 June.
   O. a. naevius.

Great Horned Owl, *Bubo virginianus* **(Gmelin)***   GHOW
Common permanent resident in the south; becoming rare in heavily forested regions, north nearly to Hudson Bay. Egg dates: 31 January to 21 April.
   B. v. virginianus breeds in southern Ontario, intergrading with the following subspecies east of Lake Superior.
   B. v. scalariventris breeds throughout most of northern Ontario; wanders to the south in winter; intergrades with *subarcticus* in the west.
   B. v. subarcticus (*wapacuthu* of AOU, 1957) breeds on the extreme western fringes of the province, wanders elsewhere.
   B. v. heterocnemis wanders, especially in the south, in winter.

Snowy Owl, *Nyctea scandiaca* **(Linnaeus)**   SNOW
Occasional, rare straggler in summer; sight records along the north coast, and at Kingston, Ottawa, and Toronto. Rare to uncommon winter resident; highest numbers about every four years. Rare transient. Principal dates in province: mid October to late April.

Northern Hawk Owl, *Surnia ulula* **(Linnaeus)***   NOHO
Rare permanent resident across the northern forested parts of the province; south, at least on occasion, to Thunder Bay and Lake Abitibi; once south to Ottawa. Occasional, rare in winter in the south (to Lake Ontario). Egg dates: unknown.
   S. u. caparoch.

Burrowing Owl, *Athene cunicularia* **(Molina)**   BUOW
Occasional, rare straggler. One specimen and seven sight records since first reported in 1940 (Baillie, 1964), mainly in the south, but also Thunder Bay. Dates in province: May; (June and July); October.
   A. c. hypugaea.

Barred Owl, *Strix varia* **Barton***   **BAOW**
Rare to uncommon permanent resident across central Ontario, north to about Sandy Lake and Cochrane, south to southern Georgian Bay and Kingston; occasional, rare south to Lake Erie. Some northern birds move south on occasion in winter. Egg dates: 4 April to 18 May.
   S. v. varia.

Great Gray Owl, *Strix nebulosa* **Forster***   **GGOW**
Rare permanent resident in wooded portions of the north; occasional south at least to Sudbury (and a breeding report from Algonquin Park); summer record from Main Duck Island (Lake Ontario). Occasional, rare (to locally uncommon) in the south in winter, mainly December to April. Egg dates: 29 April to 5 June.
   S. n. nebulosa.

Long-eared Owl, *Asio otus* **(Linnaeus)***   **LEOW**
Rare (?) summer resident throughout the province in forested areas. Rare (to locally uncommon) winter resident in the south. Egg dates: 19 March to 24 May.
   A. o. wilsonianus.

Short-eared Owl, *Asio flammeus* **(Pontoppidan)***   **SEOW**
Rare (to locally uncommon) summer resident, mainly south of the Canadian Shield and along north coast areas, but possible anywhere in the province. Locally uncommon winter resident in the south. Rare to locally uncommon migrant. Principal dates in province: early March to late October. Egg dates: 14 April to 1 August.
   A. f. flammeus.

Boreal Owl, *Aegolius funereus* **(Linnaeus)***   **BOOW**
Uncommon (but rarely encountered) permanent resident in the Boreal Forest Region of the north. Occasional, rare straggler in winter in the south. Egg dates: 14 to 27 May.
   A. f. richardsoni.

Northern Saw-whet Owl, *Aegolius acadicus* **(Gmelin)***   **NSWO**
Uncommon (but rarely encountered) summer resident across the province on the Canadian Shield, north at least to Lac Seul and Cochrane, probably somewhat farther; rare in southern agricultural areas, south to Lake Erie. Rare (to locally uncommon) winter resident in the south; occasional, rare north to Thunder Bay. Rare (to locally common) migrant. Principal dates in province: early March to late October. Egg dates: 1 April to 27 July.
   A. a. acadicus.

Lesser Nighthawk, *Chordeiles acutipennis* **(Hermann)**   **LENI**
Vagrant; one photographic record: (ROM PR 330–337) Point Pelee, 29 April 1974 (Goodwin, 1974b).

Common Nighthawk, *Chordeiles minor* **(Forster)***   **CONI**
Common summer resident throughout the south; becoming uncommon through the Boreal Forest Region and Hudson Bay Lowland; occasional, rare to the Hudson Bay coast. Common (to locally abundant) migrant. Dates in province: (early April) mid May to late September (late October). Egg dates: 26 May to 13 August.
   C. m. minor.

Common Poorwill, *Phalaenoptilus nuttallii* **(Audubon)**   **COPW**
Vagrant; one specimen: (NMC 78695) North Point, James Bay, 4 June 1982 (Godfrey, 1986).
   P. n. nuttallii (Godfrey, 1986).

Chuck-will's-widow, *Caprimulgus carolinensis* **Gmelin***   **CWWI**
Rare summer resident in the extreme south (mainly St. Williams, Rondeau Provincial Park, and Point Pelee); also recorded near Kingston. First nest reported in 1977 (Goodwin, 1977b). Dates in province: early May to July (?). Egg date: 5 June.

Whip-poor-will, *Caprimulgus vociferus* **Wilson***   **WPWI**
Uncommon (to locally common) summer resident on the Canadian Shield in the south; rare in agricultural areas, and in the north to Kenora, Nipigon, and Lake Abitibi; occasional to the latitude of Red Lake and Moosonee. Rare migrant. Dates in province: (early April) mid May to early October (early November). Egg dates: 21 May to 8 July.
   C. v. vociferus.

Chimney Swift, *Chaetura pelagica* **(Linnaeus)***   **CHSW**
Uncommon (to common) summer resident in urban areas across the south; rare in agricultural and forested areas of the south and in the north to Pickle Lake and Kapuskasing. Vagrant in winter (Goodwin, 1980a). Common migrant. Dates in province: (early April) late April to early October (early January). Egg dates: 24 May to 4 August.

Broad-billed Hummingbird, *Cynanthus latirostris* **Swainson**   **BBLH**
Vagrant; one photographic record: (ROM PR 2083–2084) Peterborough County, 16 to 26 October 1989 (Wormington and Curry, 1990; Carpentier, 1990).

Ruby-throated Hummingbird, *Archilochus colubris* **(Linnaeus)***   **RTHU**
Uncommon summer resident across the province, north to Kenora, Lake Nipigon, and Kapuskasing; occasional, rare north to Sandy Lake and Attawapiskat. Common migrant. Dates in province: (early April) mid May to mid September (mid November). Egg dates: 25 May to 2 September.

Rufous Hummingbird, *Selasphorus rufus* **(Gmelin)**   **RUHU**
Occasional, rare straggler. Three specimens and at least six sight records since the first record in 1966 (Barlow, 1967). Dates in province: June to mid December.

Belted Kingfisher, *Ceryle alcyon* **(Linnaeus)***   **BEKI**
Common summer resident in the south; becoming rare in the Boreal Forest Region, north to the Hudson Bay coast. Rare winter resident in the south (north to Sault Ste. Marie). Common migrant. Principal dates in province: early April to mid October. Egg dates: 4 May to 2 July.
    C. a. alcyon.

Lewis' Woodpecker, *Melanerpes lewis* **(Gray)**   **LEWO**
Vagrant; one photographic record: (ROM PR 553) Windsor, 10 February 1973 (Goodwin, 1973b). Two sight records: Emo, Rainy River District, 27 May 1934 (Snyder, 1938); Point Pelee, 27 October 1972 (Goodwin, 1973a).

Red-headed Woodpecker,   **RHWO**
    Melanerpes erythrocephalus **(Linnaeus)***
Rare (to locally uncommon in the extreme south) summer resident across the province, north to Kenora, Wawa, and Sudbury. Rare winter resident in the south. Rare to common migrant. Principal dates in province: early May to late September. Egg dates: 14 May to 21 July.
    M. e. erythrocephalus.

Red-bellied Woodpecker, *Melanerpes carolinus* **(Linnaeus)***   **RBWO**
Rare (to locally uncommon in the extreme south) permanent resident (north to Huron County, Durham RM, and Prince Edward County); vagrant in the north (western Rainy River District) in summer. Rare straggler north to Thunder Bay and Moosonee outside the breeding season. Egg dates: 1 to 27 May.
    M. c. zebra.

Yellow-bellied Sapsucker, *Sphyrapicus varius* **(Linnaeus)***   **YBSA**
Common summer resident across central Ontario; becoming rare (to absent in the southwest), south of the Shield, south to Lake Erie; rare north to Big Trout Lake and Fort Albany. Occasional, rare winter resident in the south (north to Owen Sound and Kingston). Common migrant. Principal dates in province: early April to mid October. Egg dates: 15 May to 12 July.
    S. v. varius.

Downy Woodpecker, *Picoides pubescens* **(Linnaeus)***   DOWO
Common permanent resident across the province; becoming rare north of Lake Nipigon to Big Trout Lake and Attawapiskat. Egg dates: 2 May to 1 July.
   P. p. medianus.

Hairy Woodpecker, *Picoides villosus* **(Linnaeus)***   HAWO
Common permanent resident across the province; becoming rare north of Lake Nipigon, to the northern Hudson Bay Lowland, and at least along major rivers almost to the Hudson Bay coast. Egg dates: 16 April to 8 June.
   P. v. septentrionalis breeds in the far north. South of Kenora and Moosonee it intergrades extensively with *P. v. villosus* throughout the rest of the north.
   P. v. villosus breeds in the south, intergrading with *P. v. septentrionalis* in the north, to Kenora and Moosonee.

Three-toed Woodpecker, *Picoides tridactylus* **(Linnaeus)***   TTWO
Rare to locally uncommon permanent resident in the north, south to Thunder Bay and Gogama; occasional, rare farther south (to Frontenac County). Wanders into the south in winter (to Lake Erie). Egg dates: 27 May to 6 June.
   P. t. bacatus.

Black-backed Woodpecker, *Picoides arcticus* **(Swainson)***   BBWO
Rare to locally uncommon permanent resident in the north, and south on the Canadian Shield to central Hastings County. Wanders south in winter (to Lake Erie). Egg dates: 18 May to 18 June.

Northern Flicker, *Colaptes auratus* **(Linnaeus)***   NOFL
Common summer resident throughout most of the province (becoming rare in the northern Hudson Bay Lowland). Rare winter resident in the south (north to Georgian Bay and Ottawa). Common migrant. Principal dates in province: early April to late October. Egg dates: 5 May to 20 July.
   C. a. luteus.
   C. a. [collaris] vagrant (ROM PR 2079–2082) at Thunder Bay, 18 November 1988 to 9 January 1989.

Pileated Woodpecker, *Dryocopus pileatus* **(Linnaeus)***   PIWO
Rare to uncommon permanent resident across the province, north to Sandy Lake and Cochrane; occasional north to Big Trout Lake and Attawapiskat; probably absent in the extreme south (Essex County). Egg dates: 4 May to 28 June.
   D. p. abieticola.

Olive-sided Flycatcher, *Contopus borealis* **(Swainson)***   OSFL
Uncommon summer resident across the forested portions of the province, south to Lake Simcoe and Brockville. Uncommon migrant. Dates in province: (early May) mid May to early September (early October). Egg dates: 6 to 24 June.

Western Wood-Pewee, *Contopus sordidulus* **Sclater**   WEWP
Vagrant; one specimen: (NMC 84372) North Point, James Bay, 20 June 1984 (Godfrey, 1986). Two records based on song: Point Pelee, 17 May 1968 (Stirrett, 1973a); and 15 May 1969 (Goodwin, 1969).
    C. s. veliei.

Eastern Wood-Pewee, *Contopus virens* **(Linnaeus)***   EAWP
Common summer resident across the south; becoming uncommon and local in the north to Woodland Caribou Provincial Park and Cochrane. Common migrant. Dates in province: (late April) mid May to mid September (late November). Egg dates: 3 June to 14 August.

Yellow-bellied Flycatcher,   YBFL
    Empidonax flaviventris **(Baird and Baird)***
Uncommon summer resident across the north; becoming rare in the south to southern Georgian Bay and Brockville. Uncommon migrant. Dates in province: (late April) mid May to mid September (late October). Egg dates: 8 June to 20 July.

Acadian Flycatcher, *Empidonax virescens* **(Vieillot)***   ACFL
Rare summer resident in the Deciduous Forest Region (north to Durham RM); sight records to Prince Edward County. Rare migrant. Dates in province: (late April) mid May to late August (late September). Egg dates: 10 June to 30 July.

Alder Flycatcher, *Empidonax alnorum* **Brewster***   ALFL
Common summer resident throughout most of the province; becoming uncommon and local in the extreme south. Common migrant. Dates in province: (late April) mid May to early September (early October). Egg dates: 15 June to 22 July.

Willow Flycatcher, *Empidonax traillii* **(Audubon)***   WIFL
Locally common summer resident in southern agricultural areas; becoming rare north to Sudbury. Uncommon migrant; wandering north to Rainy River and Manitouwadge. Dates in province: (early May) mid May to early September (late September). Egg dates: 13 June to 20 July.

Least Flycatcher, *Empidonax minimus* **(Baird and Baird)***   **LEFL**
Common to locally abundant summer resident across the province, north to Big Trout Lake and Fort Albany (rare and local to Fort Severn in the west). Absent from intensively farmed areas, especially Essex County. Common migrant. Dates in province: (late April) early May to mid September (late October). Egg dates: 27 May to 25 July.

Gray Flycatcher, *Empidonax wrightii* **Baird**   **GRFL**
Vagrant; one bird banded and photographed: (ROM PR 1244–1254) Toronto, 11 September 1981 (James, 1982).

[Western Flycatcher, *Empidonax difficilis* **Baird**   **WEFL**
The record mentioned in James, McLaren, and Barlow (1976) appears to have been a mistake in the source publication. No records exist.]

Eastern Phoebe, *Sayornis phoebe* **(Latham)***   **EAPH**
Common summer resident across most of the south; becoming local and rare in heavily forested areas, north to Pickle Lake and Timmins. Occasional, rare in winter in the south (north to Toronto). Common migrant. Principal dates in province: early April to mid October. Egg dates: 6 April to 4 August.

Say's Phoebe, *Sayornis saya* **(Bonaparte)**   **SAPH**
Occasional, rare straggler, north to James Bay, any month of the year but mainly autumn to early winter. First reported in 1948 (Baillie, 1964); at least five reports in the past decade.
    S. s. saya.

Vermilion Flycatcher, *Pyrocephalus rubinus* **(Boddaert)**   **VEFL**
Vagrant; one specimen: (ROM 76565) Toronto, 1 November 1949 (Baillie, 1957). One sight record: Renfrew County, 25 July 1972 (Goodwin, 1974b).
    P. r. mexicanus.

Ash-throated Flycatcher, *Myiarchus cinerascens* **(Lawrence)**   **ATFL**
Vagrant; three sight records: Point Pelee, 24 to 25 November 1962 (Stirrett, 1973b); Whitby, 29 October 1982 (Mountjoy and McRae, 1983); Prince Edward Point, 7 November 1982 (Weir, 1983a). [Toronto, 20 May 1989 (Weir, 1989d) to be verified.]

Great Crested Flycatcher, *Myiarchus crinitus* **(Linnaeus)***   **GCFL**
Common summer resident in the south; uncommon to locally rare in the north to Red Lake and Timmins. Common migrant; wandering north to Moosonee. Dates in province: (mid April) early May to early September (late October). Egg dates: 23 May to 10 July.
    M. c. boreus.

Sulphur-bellied Flycatcher, *Myiodynastes luteiventris* **Sclater**   **SBFL**
Vagrant; one photographic record: (ROM PR 1756–1760, 1862–1864) Presqu'ile Provincial Park, 28 September to 1 October 1986 (Gawn, 1987).

Cassin's Kingbird, *Tyrannus vociferans* **Swainson**   **CAKI**
Vagrant; one specimen: (ROM 81283) Algonquin Park, 4 June 1953 (Snyder, 1954). Two sight records: Point Pelee, 16 September 1963 (Woodford, 1964); Ottawa, 19 September to 8 October 1970 (Goodwin, 1971b).
   T. v. vociferans.

Western Kingbird, *Tyrannus verticalis* **Say***   **WEKI**
Occasional, rare in summer in the south (north to Manitoulin Island and Kingston), and the west (Rainy River and Kenora). Breeding reported from Kent County in 1943 (MacFayden, 1945), and verified in Rainy River District 1987 to 1989 (ROM PR 2029–2033; Carpentier, 1987). Rare migrant; wandering north to James Bay. Dates in province: (mid May) late May to mid September (mid November). Egg date: 20 June.

Eastern Kingbird, *Tyrannus tyrannus* **(Linnaeus)***   **EAKI**
Common to locally abundant summer resident across the south; uncommon in the north to Red Lake and Moosonee; occasional, rare north to Winisk and Cape Henrietta Maria. Common migrant. Dates in province: (early April) early May to mid September (mid October). Egg dates: 16 May to 5 August.

Gray Kingbird, *Tyrannus dominicensis* **(Gmelin)**   **GRAK**
Vagrant; three sight records: Kingston, 29 October 1970 (Hughes, 1971; Quilliam, 1973); Ottawa, 31 October 1982 (James, 1984b); Point Pelee, 26 July 1986 (Wormington, 1987).
   [*T. d. dominicensis.*]

Scissor-tailed Flycatcher, *Tyrannus forficatus* **(Gmelin)**   **STFL**
Rare straggler, mainly in the south. Records north to Big Trout Lake and Sudbury are usually associated with spring and autumn migration periods. Dates in province: late April to mid November.

Fork-tailed Flycatcher, *Tyrannus savana* **Vieillot**   **FTFL**
Vagrant; one photographic record: (ROM PR 940–941) Dorion, 28 to 30 October 1977 (Goodwin, 1977a).

Horned Lark, *Eremophila alpestris* **(Linnaeus)***   **HOLA**
Common summer resident on the north coast and in southern agricultural areas; rare and local on the Canadian Shield in southern Ontario, in farming areas from Sudbury to North Bay, and north to the Clay Belt, also western Rainy River District. Uncommon winter resident in the south (north to Mani-

toulin Island). Common to abundant migrant, many remaining into late autumn and returning by mid winter. Egg dates: 23 March to 20 July.

E. a. alpestris intergrades extensively with the following subspecies along the north coast.

E. a. hoyti intergrades with the former subspecies along the north coast.

E. a. praticola breeds in Rainy River District, in southern Ontario, and north to farming areas of the Clay Belt.

Purple Martin, *Progne subis* **(Linnaeus)***   PUMA

Locally common summer resident north to Georgian Bay and Ottawa; uncommon local to Sudbury; occasional straggler in the north to Kenora and Thunder Bay; vagrant to Fort Severn (Manning, 1952). Common to locally abundant migrant. Dates in province: (late March) mid April to mid September (late October). Egg dates: 21 May to 7 August.

P. s. subis.

Tree Swallow, *Tachycineta bicolor* **(Vieillot)***   TRES

Common to locally abundant summer resident throughout the province, except rare in the Hudson Bay Lowland. Vagrant in winter (Weir, 1985a). Common to locally abundant migrant. Dates in province: (late February) mid April to early October (late December). Egg dates: 25 April to 7 August.

Northern Rough-winged Swallow,   NRWS
   Stelgidopteryx serripennis **(Audubon)***

Locally uncommon summer resident, north to Sault Ste. Marie and North Bay; locally rare to Kenora, Thunder Bay, and Chapleau. Vagrant in winter (Goodwin, 1980a). Uncommon (to locally common) migrant. Dates in province: (mid March) early May to mid September (early January). Egg dates: 4 May to 22 July.

S. s. serripennis.

Bank Swallow, *Riparia riparia* **(Linnaeus)***   BANS

Locally common to abundant summer resident in the agricultural areas of the south; becoming rare on the Canadian Shield and in forested areas, north to the Hudson Bay coast. Common (to locally abundant) migrant. Dates in province: (mid March) early May to early September (late October). Egg dates: 4 May to 17 July.

R. r. riparia.

Cliff Swallow, *Hirundo pyrrhonota* **Vieillot***   CLSW

Locally common summer resident north to Pickle Lake and Moosonee; occasional, rare north to Hudson Bay. Common migrant. Dates in province:

(mid March) early May to early September (early November). Egg dates: 22 May to 2 August.

H. p. pyrrhonota.

Cave Swallow, *Hirundo fulva* **Vieillot**   CASW
Vagrant; one sight record: Point Pelee, 21 April 1989 (Weir, 1989d; Wormington and Curry, 1990).

Barn Swallow, *Hirundo rustica* **Linnaeus***   BARS
Abundant summer resident in the south; becoming locally uncommon in the north to Hudson Bay. Vagrant in winter (Goodwin, 1980a). Common to locally abundant migrant. Dates in province: (mid March) late April to early September (mid January). Egg dates: 10 May to 21 August.

H. r. erythrogaster.

Gray Jay, *Perisoreus canadensis* **(Linnaeus)***   GRAJ
Uncommon (to common) permanent resident across the north; becoming rare in the south on the Canadian Shield to Parry Sound, central Peterborough County, and southern Renfrew County; occasional to Ottawa. Occasional, rare south of the breeding range in winter (to Lake Erie). Egg dates: 7 March to 1 May.

P. c. canadensis.

Blue Jay, *Cyanocitta cristata* **(Linnaeus)***   BLJA
Common summer resident across the south; becoming uncommon in the north to Red Lake and Kapuskasing. Uncommon winter resident in the south, also to Thunder Bay in the north. Common to locally abundant migrant. Egg dates: 15 April to 8 July.

C. c. bromia.

Clark's Nutcracker, *Nucifraga columbiana* **(Wilson)**   CLNU
Vagrant; one photographic record: (ROM PR 365) near Dryden, 14 November 1972 to 19 June 1973 (Goodwin, 1973a). One sight record: Caribou Island, 9 May 1981 (Wormington, 1985).

Black-billed Magpie, *Pica pica* **(Linnaeus)***   BBMA
Rare permanent resident in the west (western Rainy River District). Occasional, rare elsewhere in autumn and winter. Egg date: 24 May.

P. p. hudsonia.

Jackdaw, *Corvus monedula* **Linnaeus**   JACK
Vagrant; two sight records: Whitby, 13 April 1985 (Bain, 1986); Toronto, 20 October 1985 (Coady, 1988); accepted by OBRC, but there is some doubt that these were natural occurrences (Anonymous, 1987).

American Crow, *Corvus brachyrhynchos* **Brehm***　　　　　　　　　　　**AMCR**
Common summer resident north to Kenora and Cochrane; becoming rare and local, north to Hudson Bay. Common to locally abundant winter resident in the Deciduous Forest Region; rare to locally uncommon elsewhere in the south and in the north at Thunder Bay; occasional, rare elsewhere in the north (to Atikokan and Moosonee). Common to locally abundant migrant. Egg dates: 3 March to 20 June.
　　C. b. brachyrhynchos.

Fish Crow, *Corvus ossifragus* **Wilson**　　　　　　　　　　　　　　　　**FICR**
Vagrant; three sight records: Point Pelee, 15 May 1978 (Goodwin, 1978); 21 April 1982 (James, 1983); and 20 May 1983 (James, 1984b).

Common Raven, *Corvus corax* **Linnaeus***　　　　　　　　　　　　　**CORA**
Common permanent resident across central Ontario; becoming uncommon to rare in the Hudson Bay Lowland and in the south, as far south as the Bruce Peninsula, and central Peterborough and Leeds counties. Wanders occasionally farther south (to Lake Erie). Egg dates: 10 March to 16 May.
　　C. c. principalis.

Black-capped Chickadee, *Parus atricapillus* **Linnaeus***　　　　　　　**BCCH**
Common permanent resident across the province, north to Big Trout Lake and Moosonee, rare to Little Sachigo Lake and Attawapiskat. Occasionally irruptive. Egg dates: 26 April to 14 July.
　　P. a. atricapillus.

Carolina Chickadee, *Parus carolinensis* **Audubon**　　　　　　　　　　**CACH**
Vagrant; one specimen: (ROM 28494) Long Point, 18 May 1983 (James, 1984b). Two other records: a sight report at Toronto, 10 April 1914 (Harrington, 1915); and a report based on song at Rondeau Provincial Park, 3 July 1960 (Jarvis, 1965).
　　P. c. extimus (Parkes, 1988).

[Mountain Chickadee, *Parus gambeli* **Ridgway**　　　　　　　　　　　**MOCH**
No description exists for a sight record at Port Credit, Peel RM, April 1963 (Woodford, 1963); it may have been an aberrantly plumaged Black-capped Chickadee, as was a bird (ROM PR 1673) in Simcoe County, December 1984, first identified as this species.]

Boreal Chickadee, *Parus hudsonicus* **Forster***　　　　　　　　　　　**BOCH**
Uncommon permanent resident in coniferous forests across the province, south to Sudbury; rare (to uncommon) in the Algonquin highlands. Occa-

sional, rare south of the breeding range in winter (to Point Pelee). Egg dates: 22 May to 17 June.
P. h. hudsonicus.

Tufted Titmouse, *Parus bicolor* Linnaeus*   **TUTI**
Locally rare permanent resident in the Deciduous Forest Region; wanders north to Lake Simcoe. Wanders outside the breeding season, north to Ottawa. Egg dates: 15 May to 15 June.
P. b. bicolor.

Red-breasted Nuthatch, *Sitta canadensis* Linnaeus*   **RBNU**
Uncommon summer resident across the province, north to Big Trout Lake and Moosonee; rare and local south of the Canadian Shield. Rare (to locally common) winter resident in the south; occasional, rare in the north to Thunder Bay and Manitouwadge. Uncommon to locally common (irruptive) migrant. Egg dates: 9 May to 8 June.

White-breasted Nuthatch, *Sitta carolinensis* Latham*   **WBNU**
Uncommon permanent resident in the south; occasional, rare in the north to Kenora, Thunder Bay, and Timiskaming District. Irruptive, becoming locally uncommon during movements. Egg dates: 28 April to 29 May.
S. c. cookei.

Brown Creeper, *Certhia americana* Bonaparte*   **BRCR**
Uncommon summer resident across the province, north to Big Trout Lake and Moosonee; occasional, rare north to Sutton Lake; rare and local in southern agricultural areas. Rare winter resident in the south, and in the west (to Thunder Bay). Common (to locally abundant) migrant. Egg dates: 23 April to 13 July.
C. a. americana.

Rock Wren, *Salpinctes obsoletus* (Say)   **ROWR**
Vagrant; two specimens: (ROM 94722) Niagara RM, 7 December 1964 (Baillie, 1964); (ROM 115958) Ear Falls, Kenora District, October 1972 (Howe, 1973). One photographic record: Toronto, 12 February to 5 March 1989 (Weir, 1989c).
S. o. obsoletus.

Carolina Wren, *Thryothorus ludovicianus* (Latham)*   **CARW**
Rare permanent resident in the Deciduous Forest Region; occasional, nesting north to Simcoe County. Wanders on occasion, in spring and late summer into winter, north to Sault Ste. Marie and North Bay; vagrant in the north at Marathon, Longlac, and New Liskeard. Egg dates: 5 April to 8 August.
T. l. ludovicianus.

Bewick's Wren, *Thryomanes bewickii* **(Audubon)***   **BEWR**
Former rare breeder at Point Pelee (1950, 1956, and 1957). Vagrant in winter (Baillie, 1950b). Rare spring transient and occasional autumn straggler. First recorded in 1898 (Saunders and Dale, 1933); at least six reports in the past decade. Principal dates in province: late April to late October. Egg dates: 9 to 24 May.
    T. b. altus?

House Wren, *Troglodytes aedon* **Vieillot***   **HOWR**
Common summer resident north to Georgian Bay and Ottawa; locally rare and occasional, north to Sioux Lookout and Timmins; vagrant to southern James Bay. Uncommon migrant. Dates in province: (late March) late April to late September (late December). Egg dates: 12 May to 1 September.
    T. a. baldwini breeds in the south, intergrading with the following subspecies in southern Algoma and Sudbury districts.
    T. a. parkmanii breeds in northern Ontario.

Winter Wren, *Troglodytes troglodytes* **(Linnaeus)***   **WIWR**
Common summer resident across the province; becoming locally rare in the northern Hudson Bay Lowland, south of Georgian Bay and the Canadian Shield, south to Rondeau Provincial Park. Rare winter resident in the south (north to Kingston). Common migrant. Principal dates in province: mid April to mid October. Egg dates: 31 May to 12 July.
    T. t. hiemalis.

Sedge Wren, *Cistothorus platensis* **(Latham)***   **SEWR**
Rare to locally uncommon summer resident across the province, north to Kenora, Sault Ste. Marie, and Sudbury; occasional, rare in the west north to Sandy Lake; vagrant to James Bay. Rare migrant. Dates in province: (mid April) mid May to late September (early November). Egg dates: 5 June to 22 July.
    C. p. stellaris.

Marsh Wren, *Cistothorus palustris* **(Wilson)***   **MAWR**
Locally common summer resident south of the Canadian Shield, uncommon and sparsely distributed in the rest of the south; rare in the Lake of the Woods to Lac Seul area; occasional near Thunder Bay; an isolated small colony reported from the North Point area (James Bay). Occasional, rare winter resident in the extreme south. Uncommon migrant. Principal dates in province: late April to early October. Egg dates: 9 May to 15 August.
    C. p. dissaeptus.

Golden-crowned Kinglet, *Regulus satrapa* **Lichtenstein***   **GCKI**
Common summer resident across the central forested portions of the province; becoming rare and local in the northern Hudson Bay Lowland and

south of the Canadian Shield. Uncommon winter resident in the south. Common (to locally abundant) migrant. Egg date: 30 May.
R. s. satrapa.

Ruby-crowned Kinglet, *Regulus calendula* **(Linnaeus)*** RCKI
Common summer resident across the north and on the Canadian Shield in the south; occasional, rare south of the Shield. Rare winter resident in the south (north to Durham RM). Abundant migrant. Principal dates in the province: early April to early November. Egg dates: 2 to 23 June.
R. c. calendula.

Blue-gray Gnatcatcher, *Polioptila caerulea* **(Linnaeus)*** BGGN
Uncommon summer resident in the Deciduous Forest Region; occasional, rare north to Bruce and Simcoe counties, and Ottawa-Carleton RM; vagrant to Manitoulin Island, and in the north to Rainy River. Uncommon migrant; vagrant north to Hudson Bay. Dates in province: (late March) early May to mid September (late December). Egg dates: 19 May to 1 July.
P. c. caerulea.

Siberian Rubythroat, *Luscinia calliope* **(Pallas)** SIRU
Vagrant; one specimen: (ROM 148368) Hornby, Halton RM, 26 December 1983 (Brewer, Lane, and Wernaart, 1984).
L. c. camtschatkensis.

Northern Wheatear, *Oenanthe oenanthe* **(Linnaeus)** NOWH
Occasional summer records from Cape Henrietta Maria and Winisk. Rare transient throughout the province, mainly in the autumn. Dates in province: mid March to early June (early July); late August to mid October.
O. o. leucorhoa.

Eastern Bluebird, *Sialia sialis* **(Linnaeus)*** EABL
Locally common summer resident in agricultural areas of the south; rare (to locally common) in the north to Lac Seul and Cochrane; formerly and still possible north to Favourable Lake and Moosonee. Rare (to locally uncommon) winter resident in the south (north to Kingston). Uncommon migrant. Principal dates in province: early April to late October. Egg dates: 10 April to 2 September.
S. s. sialis.

Mountain Bluebird, *Sialia currucoides* **(Bechstein)*** MOBL
Unexpected summer resident; male paired with an Eastern Bluebird in 1985 and 1986 in Elgin County (Weir, 1985b, 1986c; but see also Wormington and Curry, 1990). Occasional, rare migrant in the south and at Atikokan, Thunder Bay, and Rossport in the north, usually in autumn. Vagrant in winter (Weir, 1989d). First reported in 1962 (Wyett, 1966); at least 12 reports

in the past decade. Dates in province: any month; most frequently April and December. Egg dates: 5 to 25 May.

Townsend's Solitaire, *Myadestes townsendi* **(Audubon)** TOSO
Occasional, rare straggler, mainly in the south, but north to Thunder Bay, Marathon, and Cochrane. First reported in 1960 (Speirs, 1960c); at least nine reports in the past decade. Dates in province: mid September to late April.
  M. t. townsendi.

Veery, *Catharus fuscescens* **(Stephens)*** VEER
Common summer resident in the south; becoming uncommon in the north to Sioux Lookout and Cochrane. Common migrant. Dates in province: (mid April) early May to mid September (late October). Egg dates: 3 May to 17 July.
  C. f. salicicola breeds in the west, east to Thunder Bay; also appears at Sault Ste. Marie and on Manitoulin Island.
  C. f. fuscescens breeds in southern Ontario and the eastern part of northern Ontario, west to Lake Nipigon.

Gray-cheeked Thrush, *Catharus minimus* **(Lafresnaye)*** GCTH
Locally rare and likely irregular summer resident near Hudson Bay (Fort Severn and Hawley Lake), and possibly throughout the northern Hudson Bay Lowland. Common (to locally abundant) migrant. Dates in province: (mid April) mid May to early October (late December). Egg dates: 2 to 8 July.
  C. m. bicknelli rare migrant in the south.
  C. m. aliciae breeds.

Swainson's Thrush, *Catharus ustulatus* **(Nuttall)*** SWTH
Common summer resident across the province, south to the Bruce Peninsula and central Peterborough County; occasional, rare to Wellington and southern Frontenac counties. Common (to locally abundant) migrant. Dates in province: (early March) early May to late September (late December). Egg dates: 28 May to 29 July.
  C. u. swainsoni.

Hermit Thrush, *Catharus guttatus* **(Pallas)*** HETH
Common summer resident in the north; in the south, confined mainly to the Canadian Shield, but rare and local to Bruce County, Durham RM, and the St. Lawrence River, also Haldimand-Norfolk RM. Rare winter resident in the south (north to Ottawa). Common migrant. Principal dates in province: mid April to late October. Egg dates: 14 May to 15 August.
  C. g. faxoni.

Wood Thrush, *Hylocichla mustelina* **(Gmelin)*** WOTH
Common summer resident north to about Georgian Bay and Ottawa; becoming rare north to Sault Ste. Marie and Sudbury; summer records north to Rainy River, Thunder Bay, Wawa, and Cochrane; vagrant on James Bay (North Point). Common migrant. Dates in province: (early April) early May to late September (early December). Egg dates: 4 May to 27 July.

Eurasian Blackbird, *Turdus merula* **Linnaeus** EUBL
Vagrant; one sight record: Erieau, Kent County, 12 April 1981 (James, 1984b).

Fieldfare, *Turdus pilaris* **Linnaeus** FIEL
Vagrant; two photographic records: (ROM PR 400–404) Long Point, 24 May 1975 (Hussel and Porter, 1977); (ROM PR 1277–1281) Toronto, 1 January to March 1981 (Goodwin, 1981). One sight record: Ottawa, 8 January 1967 (MacKenzie, 1968).

American Robin, *Turdus migratorius* **Linnaeus*** AMRO
Abundant summer resident in residential and agricultural areas; becoming uncommon in forested areas throughout the province. Rare (to locally uncommon) winter resident in the south; occasional, rare in the north to Thunder Bay and Kirkland Lake. Abundant migrant. Principal dates in province: mid March to mid November. Egg dates: 8 April to 15 August.
T. m. migratorius.

Varied Thrush, *Ixoreus naevius* **(Gmelin)** VATH
Vagrant in summer (Weir, 1988c). Rare transient and winter resident in the south; occasional in the north to Dryden, the Thunder Bay area, and Timmins. First reported in 1963 (Baillie, 1964). Dates in province: late September to mid April (25 June).
I. n. meruloides.

Gray Catbird, *Dumetella carolinensis* **(Linnaeus)*** GRCA
Common summer resident in the south; becoming locally rare north to Kenora and Timmins; vagrant to southern James Bay and Big Trout Lake. Occasional, rare in winter in the south (north to Toronto). Common migrant. Principal dates in province: early May to late September. Egg dates: 2 May to 18 August.

Northern Mockingbird, *Mimus polyglottos* **(Linnaeus)*** NOMO
Uncommon permanent resident in the Niagara Peninsula; rare and irregular elsewhere in the south; occasional north to Thunder Bay and Moosonee; vagrant to Big Trout Lake. Winters in the south; occasional north to

Thunder Bay and Sault Ste. Marie. Partially migratory. Egg dates: 23 May to 8 August.
M. p. polyglottos.

Sage Thrasher, *Oreoscoptes montanus* **(Townsend)** SATH
Occasional, rare straggler at Atikokan, Thunder Bay, Manitoulin District, and near Lake Erie. First reported in 1965 (Baillie, 1965); at least five reports in the past decade. Dates in province: late April to mid October.

Brown Thrasher, *Toxostoma rufum* **(Linnaeus)*** BRTH
Common summer resident in the south; becoming local and rare in the north to Kenora and Timmins; vagrant to Moosonee. Rare winter resident, mainly in the south, but occasionally north to Thunder Bay and Kirkland Lake. Common migrant. Principal dates in province: mid April to mid October. Egg dates: 20 April to 20 July.
T. r. rufum.

American Pipit, *Anthus rubescens* **(Tunstall)*** AMPI
Common summer resident along the coast of Hudson Bay. Occasional, rare winter resident in the south (north to Peterborough). Common autumn and rare (to locally common) spring migrant. Principal dates in province: late April to early November. Egg dates: 29 June to 2 July.

Sprague's Pipit, *Anthus spragueii* **(Audubon)** SPPI
Vagrant; two song recordings: western Rainy River District, 12 July 1980 (Lamey, 1981), and 2 June 1990.

Bohemian Waxwing, *Bombycilla garrulus* **(Linnaeus)*** BOWA
Locally rare summer resident in the northern Hudson Bay Lowland. Rare to abundant erratic, autumn to spring, south to Lake Erie. Egg dates: unknown.
B. g. pallidiceps.

Cedar Waxwing, *Bombycilla cedrorum* **Vieillot*** CEDW
Common summer resident across the province, north at least to Big Trout Lake and Attawapiskat, probably rare to the north coast in the west and to Sutton Lake in the east. Common winter erratic in the south; occasional in the north to northern Lake Superior. Common to locally abundant migrant. Egg dates: 21 May to 16 September.

Phainopepla, *Phainopepla nitens* **(Swainson)** PHAI
Vagrant; one photographic record: (ROM PR 411–415) Elgin County, 27 December 1975 to 17 January 1976 (Goodwin, 1976a); and the same bird (?) seen at London, 29 February to 1 March 1976 (Goodwin, 1976a).

Northern Shrike, *Lanius excubitor* **Linnaeus*** NSHR
Rare summer resident in the northern Hudson Bay Lowland, and southward, in the east at least, to Moosonee. Rare straggler and winter resident throughout the province. Egg dates: unknown.
L. e. borealis.

Loggerhead Shrike, *Lanius ludovicianus* **Linnaeus*** LOSH
Rare summer resident in the south. Declining in recent years and now absent from large areas, but concentrated mainly near the southern edge of the Canadian Shield. Formerly rare breeder at Rainy River and Thunder Bay; now a rare straggler. Rare migrant. Dates in province: (late February) early April to mid September (late December). Egg dates: 1 April to 5 August.
L. l. migrans.

European Starling, *Sturnus vulgaris* **Linnaeus*** EUST
Abundant permanent resident in the south; becoming a rare and local summer resident north to Winisk and Cape Henrietta Maria. Partially migratory; wintering north to Kenora and Cochrane; occasional to Moosonee. Egg dates: 26 March to 19 July. Introduced to North America (1890); first reported in Ontario in 1914; first nesting in 1922.
S. v. vulgaris.

White-eyed Vireo, *Vireo griseus* **(Boddaert)*** WEVI
Rare (to locally uncommon) summer resident near Lake Erie; summer sightings north to Manitoulin Island and Ottawa. Formerly and probably erroneously reported breeding at Toronto in 1898 (Macoun and Macoun, 1909), and Niagara RM (Beardslee and Mitchell, 1965); first confirmed nesting in 1971 (Rayner, 1988). Rare migrant; vagrant in the north (at Marathon). Dates in province: (early April) early May to mid October (late December). Egg dates: 7 to 11 June.
V. g. noveboracensis.

Bell's Vireo, *Vireo bellii* **Audubon** BEVI
Occasional, rare straggler in spring in the extreme south (Point Pelee); sight records north to Presqu'ile Park, all in May. First reported in 1940 (Baillie, 1964); at least five reports in the past decade.
V. b. bellii.

Solitary Vireo, *Vireo solitarius* **(Wilson)*** SOVI
Common summer resident across central Ontario; becoming uncommon in the boreal forest, north to Big Trout Lake and Fort Albany, and in the south to Muskoka DM and central Peterborough County; occasional, rare to within 100 km of the north coast, and south to Haldimand-Norfolk RM (regular at St. Williams). Uncommon migrant. Dates in province: (late

March) early May to mid October (mid December). Egg dates: 18 May to 5 August.
V. s. solitarius.

Yellow-throated Vireo, *Vireo flavifrons* **Vieillot***   **YTVI**
Uncommon (to locally common) summer resident in the south (north to Bruce County and Ottawa); occasional, rare to Sault Ste. Marie, Manitoulin Island, and southern Renfrew County, and in the north in western Rainy River District. Rare migrant. Dates in province: (early April) mid May to mid September (mid November). Egg dates: 21 May to 15 July.

Warbling Vireo, *Vireo gilvus* **(Vieillot)***   **WAVI**
Common summer resident in the south (except largely absent from the Algonquin Highlands), and in the north in western Rainy River District; becoming occasional, rare north to Lac Seul and Cochrane. Uncommon migrant. Dates in province: (mid April) early May to mid September (mid November). Egg dates: 20 May to 5 July.
V. g. gilvus.

Philadelphia Vireo, *Vireo philadelphicus* **(Cassin)***   **PHVI**
Uncommon to common summer resident across the north, at least as far north as Big Trout Lake and Attawapiskat (possibly to Fort Severn in the west); becoming occasional, rare in the south as far south as the Bruce Peninsula, and central Peterborough and northern Leeds counties. Uncommon migrant. Dates in province: (late April) mid May to mid September (early November). Egg dates: 13 June to 25 July.

Red-eyed Vireo, *Vireo olivaceus* **(Linnaeus)***   **REVI**
Abundant summer resident across the province, north to Big Trout Lake and Attawapiskat, probably almost to Fort Severn along major rivers in the west. Common migrant. Dates in province: (early April) mid May to early October (early November). Egg dates: 26 May to 10 August.

Blue-winged Warbler, *Vermivora pinus* **(Linnaeus)***   **BWWA**
Rare to locally uncommon summer resident in the south (north to Georgian Bay and Kingston). Rare migrant; vagrant in the north (Marathon and Moosonee). Dates in province: (mid April) early May to early September (early December). Egg dates: 4 to 18 June.
    (Brewster's and Lawrence's Warblers, hybrid offspring of this and the following species, are of rare occurrence in the south; Brewster's is also vagrant in the north to Caribou Island.)

Golden-winged Warbler, *Vermivora chrysoptera* **(Linnaeus)***   **GWWA**
Uncommon (to locally common) summer resident in the south; spring and summer records north to Rainy River and Thunder Bay. Rare migrant.

Dates in province: (late April) early May to early September (late October).
Egg dates: 20 May to 3 July.

Tennessee Warbler, *Vermivora peregrina* **(Wilson)*** TEWA
Common to locally abundant summer resident across the north; becoming uncommon in the south in the Algonquin Highlands; occasional, rare on the Bruce Peninsula and on the Canadian Shield south to Kingston. Common migrant. Dates in province: (mid April) mid May to late September (mid November). Egg dates: 8 June to 21 July.

Orange-crowned Warbler, *Vermivora celata* **(Say)*** OCWA
Uncommon summer resident in the northern third of the province; becoming rare and local, south to Kenora and Timmins. Vagrant in winter (Weir, 1988a). Rare migrant. Dates in province: (late April) early May to mid October (late January). Egg date: 14 June.
  V. c. celata.

Nashville Warbler, *Vermivora ruficapilla* **(Wilson)*** NAWA
Common summer resident across the province, north to Sandy Lake and Moosonee (occasional, rare to Fort Severn and Winisk); becoming rare south of the Canadian Shield, south to Haldimand-Norfolk RM. Vagrant in winter (Goodwin, 1968). Common migrant. Dates in province: (early April) early May to early October (early January). Egg dates: 24 May to 21 July.
  V. r. ruficapilla.

Virginia's Warbler, *Vermivora virginiae* **(Baird)** VIWA
Vagrant; one specimen: (NMC 41430) Point Pelee, 16 May 1958 (Dow, 1962). Two photographic records: (ROM PR 428–431) Pelee Island, 9 to 11 May 1974 (Goodwin, 1974b); (ROM PR 1815) Point Pelee, 3 to 4 May 1975 (Goodwin, 1975).

Northern Parula, *Parula americana* **(Linnaeus)*** NOPA
Rare to uncommon local summer resident across the province, north to Red Lake, Lake Nipigon, and Timmins; south to Bruce and Peterborough counties; sight records north to Big Trout Lake and south to Kent County. Rare (to uncommon) migrant. Dates in province: (early April) mid May to late September (late November). Egg dates: unknown.

Yellow Warbler, *Dendroica petechia* **(Linnaeus)*** YWAR
Locally common summer resident throughout the province. Vagrant in winter (Goodwin, 1980a). Common migrant. Dates in province: (mid April) early May to late August (early January). Egg dates: 15 May to 17 July.
  D. p. amnicola breeds in the north, intergrading with *aestiva* in the south.
  D. p. aestiva intergrades with *amnicola* in the south.

Chestnut-sided Warbler, *Dendroica pensylvanica* **(Linnaeus)***   **CSWA**
Common summer resident on the Canadian Shield, north to Kenora and Kapuskasing; occasional, rare north to Ney Lake and Moosonee; rare to absent in southern agricultural areas. Common migrant. Dates in province: (mid April) early May to late September (late November). Egg dates: 25 May to 22 July.

Magnolia Warbler, *Dendroica magnolia* **(Wilson)***   **MAWA**
Common summer resident across the province, north to Big Trout Lake and Fort Albany, south to southern Georgian Bay and Ottawa; occasional, rare north to Sutton Lake and near Fort Severn, south to Haldimand-Norfolk RM. Abundant migrant. Dates in province: (mid April) early May to late September (early December). Egg dates: 29 May to 17 July.

Cape May Warbler, *Dendroica tigrina* **(Gmelin)***   **CMWA**
Locally common summer resident across the province, north to Big Trout Lake and Moosonee, south to the Algonquin Park area; occasional, rare on Manitoulin Island and the Bruce Peninsula. Uncommon to common migrant. Dates in province: (late April) early May to late September (early December). Egg date: 12 June.

Black-throated Blue Warbler, *Dendroica caerulescens* **(Gmelin)***   **BTBW**
Common summer resident north to Wawa and Timmins, south to southern Georgian Bay and Ottawa; becoming rare, north to Kenora, Red Lake, and Cochrane, south to York RM and Kingston. Vagrant in winter (Bennett, 1989; Weir, 1989c). Common migrant. Dates in province: (late April) early May to early October (mid February). Egg dates: 28 May to 24 July.
    D. c. caerulescens.

Yellow-rumped Warbler, *Dendroica coronata* **(Linnaeus)***   **YRWA**
Common summer resident across the province, south to southern Georgian Bay and Ottawa, rarely south to Haldimand-Norfolk RM. Rare winter resident in the south (north to Peterborough and Ottawa). Common (to locally abundant) migrant. Principal dates in province: late April to late October. Egg dates: 24 May to 29 July.
    D. c. coronata breeds.
    D. c. memorabilis occasional, rare straggler at Thunder Bay and in the south (Point Pelee to Port Hope).

Black-throated Gray Warbler, *Dendroica nigrescens* **(Townsend)**   **BTYW**
Occasional, rare straggler in the south (north to Toronto). First reported in 1952 (Baillie, 1957); at least five reports since 1978. Dates in province: early May to mid June; mid September to early January.

Townsend's Warbler, *Dendroica townsendi* **(Townsend)** TOWA
Occasional, rare straggler in the south (north to Manitoulin Island). First photographed in 1972 (ROM PR 440–442, 1822; Goodwin, 1972a); one earlier (Stirrett, 1973a) and four subsequent sight records. All occurrences in late April and May.

Hermit Warbler, *Dendroica occidentalis* **(Townsend)** HEWA
Vagrant; one specimen record: (NMC 78708) Bath, Lennox and Addington County, 10 September 1978. One photographic record: (ROM PR 1302–1303, 1344) Point Pelee, 2 to 7 May 1981. Two sight records: Toronto, 30 April 1984; Manitoulin Island, 23 May 1989.

Black-throated Green Warbler, *Dendroica virens* **(Gmelin)*** BTNW
Common summer resident across the province, north to Sandy Lake and Moosonee; becoming rare south of the Canadian Shield to Haldimand-Norfolk RM and Rondeau Provincial Park. Common migrant. Dates in province: (late March) early May to early October (mid December). Egg dates: 5 June to 9 August.
   D. v. virens.

Blackburnian Warbler, *Dendroica fusca* **(Müller)*** BLBW
Common summer resident across the province, north to Sandy Lake and Cochrane (sight records to Fort Severn); rare south of the Canadian Shield to Elgin County. Vagrant in winter (Weir, 1982). Common migrant. Dates in province: (early April) early May to late September (early January). Egg dates: 1 June to 5 July.

Yellow-throated Warbler, *Dendroica dominica* **(Linnaeus)** YTWA
Rare straggler in the south; vagrant in the north (Moosonee). First reported in 1944 (Sheppard, 1944); more than 30 reports in the past decade. Dates in province: mid April to late May; late September to early January.
   D. d. albilora rare straggler.
   D. d. dominica one record (McRae and Hutchison, 1983).

Pine Warbler, *Dendroica pinus* **(Wilson)*** PIWA
Uncommon to locally common summer resident in the south; occasional, rare in the north to Kenora, Thunder Bay, and Wawa. Occasional, rare winter resident in the south. Rare migrant; wanders north to Moosonee. Principal dates in province: late April to late September. Egg dates: 7 June to 1 July.
   D. p. pinus.

Kirtland's Warbler, *Dendroica kirtlandii* **(Baird)***　　　　　　　　**KIWA**
Former locally uncommon summer resident (Harrington, 1939); one accepted breeding record (Speirs, 1985). Now, occasional, rare straggler and (nonbreeding?) summer resident in the south (north to Georgian Bay and Renfrew County); vagrant in the north (Minaki, Kenora District) in summer. Dates in province: early May to early October. Egg dates: unknown.

Prairie Warbler, *Dendroica discolor* **(Vieillot)***　　　　　　　　**PRAW**
Rare to locally uncommon summer resident in the south (north to Parry Sound District and Frontenac County); summer sight records to Lake Nipissing. Rare migrant. Dates in province: (mid April) early May to mid September (mid October). Egg dates: 17 May to 8 July.
   D. d. discolor.

Palm Warbler, *Dendroica palmarum* **(Gmelin)***　　　　　　　　**PAWA**
Uncommon summer resident across the northern Hudson Bay Lowland; becoming rare south to northern Lake Superior; becoming occasional, rare south of Wawa and Timmins, south to the Bruce Peninsula and Ottawa. Vagrant in winter (Kelley, 1978). Common migrant. Dates in province: (early March) late April to mid October (early January). Egg dates: 25 May to 6 July.
   D. p. hypochrysea breeds in the southeast (Ottawa), rare migrant west to Point Pelee.
   D. p. palmarum breeds elsewhere across the province.

Bay-breasted Warbler, *Dendroica castanea* **(Wilson)***　　　　　　　　**BBWA**
Common summer resident across the north, as far north as Sandy Lake and Fort Albany; uncommon in the Algonquin Highlands; occasional, rare north to near Fort Severn and Attawapiskat, south to Muskoka DM and Renfrew County. Common migrant. Dates in province: (mid April) early May to late September (early December). Egg dates: 13 June to 3 July.

Blackpoll Warbler, *Dendroica striata* **(Forster)***　　　　　　　　**BLPW**
Common summer resident in the northern Hudson Bay Lowland and south to Big Trout Lake and Moosonee; occasional, rare south to Kesagami Lake. Common migrant. Dates in province: (late April) mid May to late September (early December). Egg dates: 18 June to 3 July.

Cerulean Warbler, *Dendroica cerulea* **(Wilson)***　　　　　　　　**CERW**
Rare to locally uncommon summer resident in the south, north to Bruce County and to northern Leeds and Grenville County. Rare migrant. Dates in province: (early April) early May to early September (late October). Egg dates: 24 May to 27 June.

Black-and-white Warbler, *Mniotilta varia* **(Linnaeus)*** BAWW
Common summer resident across the province, north to Big Trout Lake and Attawapiskat, but local to absent in the Deciduous Forest Region. Vagrant in winter (Goodwin, 1980a). Common migrant. Dates in province: (early February) early May to late September (late December). Egg dates: 22 May to 30 July.

American Redstart, *Setophaga ruticilla* **(Linnaeus)*** AMRE
Common summer resident across the province, north to Sandy Lake and Fort Albany. Common migrant. Dates in province: (late April) early May to early October (mid December). Egg dates: 27 May to 27 July.
 S. r. tricolora.

Prothonotary Warbler, *Protonotaria citrea* **(Boddaert)*** PROW
Rare to locally uncommon summer resident in the Deciduous Forest Region. Rare migrant; vagrant in the north (Quetico Provincial Park and Moosonee). Dates in province: (mid April) early May to mid September (early October). Egg dates: 25 May to 3 July.

Worm-eating Warbler, *Helmitheros vermivorus* **(Gmelin)** WEWA
Rare transient in the south (north to Ottawa) in spring; nonbreeding birds known to linger into July. Occasional, rare straggler in autumn. Dates in province: late April to early July; late August to late October.

Swainson's Warbler, *Limnothlypis swainsonii* **(Audubon)** SWWA
Vagrant; two sight records: Point Pelee, 22 May 1975 (Wormington, 1985), 10 to 14 May 1986 (Wormington, 1987). A 1968 Point Pelee record (Stirrett, 1973a) has no documentation.

Ovenbird, *Seiurus aurocapillus* **(Linnaeus)*** OVEN
Abundant summer resident across the province; becoming rare through the Boreal Forest Region north to Big Trout Lake and Fort Albany, and occasionally somewhat farther. Vagrant in winter (Weir, 1988a). Common to locally abundant migrant. Dates in province: (mid April) early May to late September (early January). Egg dates: 21 May to 28 July.
 S. a. aurocapillus.

Northern Waterthrush, *Seiurus noveboracensis* **(Gmelin)*** NOWA
Common summer resident throughout the province, except rare in southern agricultural areas. Common migrant. Dates in province: (mid April) early May to mid September (late December). Egg dates: 15 May to 3 July.
 S. n. notabilis.

Louisiana Waterthrush, *Seiurus motacilla* **(Vieillot)***   **LOWA**
Rare local summer resident in the Deciduous Forest Region; occasional, rare in the Kingston area; sight records north to Ottawa. Rare migrant; wanders north to Manitoulin District. Dates in province: (late March) late April to early September (early October). Egg dates: 1 June to 8 July.

Kentucky Warbler, *Oporornis formosus* **(Wilson)**   **KEWA**
Very rare summer resident 1985 to 1988, but nesting never determined (McCracken, 1988). Rare transient in spring and autumn in the south (north to Manitoulin District). Dates in province: late April to late October.

Connecticut Warbler, *Oporornis agilis* **(Wilson)***   **CONW**
Locally uncommon summer resident across central Ontario, north to Big Trout Lake and Fort Albany, south to Wawa and Cochrane; occasional, rare south to Sault Ste. Marie and Sudbury. Rare migrant. Dates in province: (early April) mid May to late September (early October). Egg dates: unknown.

Mourning Warbler, *Oporornis philadelphia* **(Wilson)***   **MOWA**
Common summer resident across the province, north to Sandy Lake and Fort Albany. Uncommon migrant. Dates in province: (late April) mid May to mid September (late October). Egg dates: 25 May to 20 July.

MacGillivray's Warbler, *Oporornis tolmiei* **(Townsend)**   **MGWA**
Vagrant; one specimen: (AMNH 507393) Hamilton, 20 May 1890 (Lanyon and Bull, 1967).
    O. t. tolmiei (?).

Common Yellowthroat, *Geothlypis trichas* **(Linnaeus)***   **COYE**
Common summer resident across the province, north to Big Trout Lake and Attawapiskat; probably occasional farther north in the west. Occasional, rare in winter in the south (Point Pelee to Toronto). Common to locally abundant migrant. Principal dates in province: early May to late September. Egg dates: 19 May to 29 July.
    G. t. brachidactylus breeds in the east, intergrading with the following subspecies west of Lake Superior.
    G. t. campicola intergrades with *brachidactylus* in the west.

Hooded Warbler, *Wilsonia citrina* **(Boddaert)***   **HOWA**
Rare to locally uncommon summer resident in the Deciduous Forest Region; also Simcoe County in 1989. Rare migrant; spring and autumn sight records north to Moosonee. Dates in province: (early April) early May to mid September (late November). Egg dates: 9 June to 27 July.

Wilson's Warbler, *Wilsonia pusilla* **(Wilson)*** WIWA
Common summer resident across the province, south to Sudbury; summer sight records south to Algonquin Park and Ottawa. Common migrant. Dates in province: (late April) mid May to late September (early January). Egg dates: 8 June to 3 July.
W. p. pusilla.

Canada Warbler, *Wilsonia canadensis* **(Linnaeus)*** CAWA
Common summer resident across central Ontario, north to Sandy Lake and Attawapiskat; rare in southern agricultural areas and absent from the extreme southwest. Uncommon migrant. Dates in province: (late April) mid May to mid September (late November). Egg dates: 26 May to 28 June.

Painted Redstart, *Myioborus pictus* **(Swainson)** PARE
Vagrant; one specimen: Durham RM, captured 9 November 1971 (Speirs and Pegg, 1972).
M. p. pictus.

Yellow-breasted Chat, *Icteria virens* **(Linnaeus)*** YBCH
Rare (to locally uncommon) summer resident in the Deciduous Forest Region (north to Peel RM); summer sightings in the Kingston area. Vagrant in winter (Fazio, Shepherd, and Woodrow, 1985). Rare migrant; spring and autumn records, north to northern Lake Superior and southern James Bay. Dates in province: (late April) early May to mid September (early January). Egg dates: 2 June to 1 July.
I. v. virens.

Summer Tanager, *Piranga rubra* **(Linnaeus)** SUTA
Vagrant in summer in the Deciduous Forest Region (Cadman, Eagles, and Helleiner, 1987). Rare migrant in the south (north to Manitoulin Island and Ottawa); vagrant in the north (to northern Lake Superior) usually in spring. Dates in province: mid April to late July; early October to mid December.
P. r. rubra.

Scarlet Tanager, *Piranga olivacea* **(Gmelin)*** SCTA
Uncommon summer resident in the south, rare in the north to Sioux Lookout and Kapuskasing; sight records to Moosonee. Vagrant in winter (Goodwin, 1979a). Uncommon migrant. Dates in province: (early February) early May to late September (early December). Egg dates: 24 May to 13 July.

Western Tanager, *Piranga ludoviciana* **(Wilson)** WETA
Occasional, rare straggler in the south; vagrant in the north (at Thunder Bay); mainly during spring migration. First reported in 1947 (Baillie, 1958); at least five reports in the past decade. Dates in province: late March to late May; late July, late August; early December to mid February.

Northern Cardinal, *Cardinalis cardinalis* **(Linnaeus)*** NOCA
Common permanent resident south of the Canadian Shield; becoming rare north to Manitoulin Island and North Bay; occasional straggler north to Rainy River, Thunder Bay, and Kirkland Lake. Egg dates: 13 April to 15 August.
 C. c. cardinalis.

Rose-breasted Grosbeak, *Pheucticus ludovicianus* **(Linnaeus)*** RBGR
Common summer resident in the south; uncommon in the north to Sioux Lookout and Cochrane; rare north to Pickle Lake and Moosonee; vagrant at Winisk. Occasional, rare winter resident in the south (north to Sudbury), also Thunder Bay. Common migrant. Principal dates in province: early May to late September. Egg dates: 10 May to 16 July.

Black-headed Grosbeak, *Pheucticus melanocephalus* **(Swainson)** BHGR
Occasional, rare straggler, north to Kenora, Thunder Bay, and Peterborough, in spring, late summer, and winter. First reported in 1949 (Baillie, 1950a); at least six reports in the past decade. Dates in province: late January to mid May; late August to mid September.

Blue Grosbeak, *Guiraca caerulea* **(Linnaeus)** BLGR
Occasional, rare straggler in the south (north to Lake Nipissing); vagrant in the north (Vermilion Bay and Rossport), usually in May. First reported in 1913 (Beardslee and Mitchell, 1965); at least 11 reports in the past decade. Dates in province: late April to late August.
 G. c. caerulea.

Lazuli Bunting, *Passerina amoena* **(Say)** LASB
Vagrant; two photographic records: (ROM PR 1282–1285) Pickle Lake, 10 May 1979; (ROM PR 2024–2028) Dryden, 15 to 18 June 1988. Sight record: Point Pelee, 23 May 1982.

Indigo Bunting, *Passerina cyanea* **(Linnaeus)*** INBU
Common summer resident in the south; becoming rare in the north to Red Lake, Nipigon, and Timmins. Vagrant in winter (resident at Thunder Bay; Weir, 1987a). Uncommon migrant; wandering in autumn north to Winisk and Moosonee. Principal dates in province: mid May to late September. Egg dates: 26 May to 15 August.

Painted Bunting, *Passerina ciris* **(Linnaeus)** PABU
Vagrant; two photographic records: (ROM PR 1118–1120) Long Point,
21 May 1978 (Nol, 1983); (ROM PR 1193–1195) Toronto, 4 December
1978 to 1 January 1979. Sight records: Lanark County, July to August 1966;
Cobourg, 25 October 1974; Middlesex County, 29 April to 1 May 1986.
While some of these reports may be of escaped captives, I feel that wild birds
probably are involved also.

Dickcissel, *Spiza americana* **(Gmelin)*** DICK
Occasional, rare (to locally uncommon) summer resident, mainly in the
Deciduous Forest Region, but also north to the Bruce Peninsula and
Toronto; vagrant in the north (western Rainy River District). Occasional,
rare straggler in winter in the south. Rare migrant; wandering north to Thunder Bay and Fort Albany. Principal dates in province: early May to late
October. Egg dates: 6 June to 3 August.

Green-tailed Towhee, *Pipilo chlorurus* **(Audubon)** GTTO
Vagrant; three photographic records: (ROM PR 548, 1897–1900) London,
30 March 1954; (ROM PR 549) Terra Cotta, Peel RM, 24 November 1956
(Baillie, 1957); (ROM PR 1730–1733) Windsor, October 1985 to April 1986
(Wormington, 1987). Sight records: Welland, March to April 1954; Whitby,
Durham RM, 11 to 12 October 1970.

Rufous-sided Towhee, *Pipilo erythrophthalmus* **(Linnaeus)*** RSTO
Uncommon summer resident north to Muskoka DM and Ottawa; rare to
Sudbury; vagrant to Fort Severn and James Bay. Rare winter resident in
the south (to Sault Ste. Marie and Sudbury), also at Geraldton, Thunder
Bay, and Marathon in the north. Uncommon migrant. Principal dates in
province: early April to late October. Egg dates: 6 May to 13 August.
  P. e. erythrophthalmus breeds.
  [*P. e. arcticus*] occasional winter straggler.
  P. e. montanus vagrant, Niagara RM, 4 to 12 December 1976.

Bachman's Sparrow, *Aimophila aestivalis* **(Lichtenstein)** BACS
Occasional, rare straggler in spring in the Lake Erie area. Dates in province:
mid April to mid May. First reported in 1917 (Saunders, 1919); last reported
in 1966.
  A. a. bachmani.

Cassin's Sparrow, *Aimophila cassinii* **(Woodhouse)** CASP
Vagrant; four photographic records: Point Pelee (ROM PR 545–547,
1608–1616) 13 May 1967 (Long, 1968), (ROM PR 1617–1621) 19 to 23 May
1984, (ROM PR 1939, 1944) 7 to 8 May 1987; Long Point (ROM PR 2046–
2047), 15 August 1987. Sight records: Marathon, 28 September 1981; Point
Pelee, 17 May 1990 (Weir, 1990a).

American Tree Sparrow, *Spizella arborea* **(Wilson)***   ATSP
Common summer resident along the coasts of Hudson Bay and northern James Bay. Common winter resident in the south; rare in the north at Thunder Bay. Common migrant; abundant along the James Bay coast. Egg dates: 21 June to 9 July.
   S. a. arborea.

Chipping Sparrow, *Spizella passerina* **(Bechstein)***   CHSP
Common summer resident throughout the province, north to Big Trout Lake and Moosonee; rare to Fort Severn and Attawapiskat. Occasional, rare winter resident in the south (north to Ottawa). Common migrant. Principal dates in province: mid April to mid October. Egg dates: 1 May to 14 August.
   S. p. boreophila breeds in the extreme west near the Manitoba border, intergrading with the following subspecies east to about Lake Nipigon.
   S. p. passerina breeds throughout the range, west to about the longitude of Lake Nipigon, intergrading with the former subspecies west to the Manitoba border.

Clay-colored Sparrow, *Spizella pallida* **(Swainson)***   CCSP
Common local summer resident in western Rainy River District; uncommon at Thunder Bay; rare and local summer resident in the south and in the north at Moosonee; with scattered summer sightings north to Fort Severn and Attawapiskat. Rare migrant. Dates in province: (late April) early May to early October (late December). Egg dates: 29 May to 20 June.

[Brewer's Sparrow, *Spizella breweri* **Cassin**   BRSP
A bird photographed at Port Stanley, Elgin County, 16 April 1980 (ROM PR 1216–1228; Goodwin, 1980b) has *not* been accepted as a provincial record (identification uncertain).]

Field Sparrow, *Spizella pusilla* **(Wilson)***   FISP
Common summer resident in the south, north to southern Georgian Bay and Ottawa; rare north to Manitoulin Island and Lake Nipissing; straggler north to the north coast. Rare in winter in the south (north to Manitoulin Island); vagrant in the north (Lake Superior). Uncommon migrant; wandering north to Hornpayne and James Bay. Principal dates in province: mid April to late October. Egg dates: 4 May to 9 August.
   S. p. pusilla.

Vesper Sparrow, *Pooecetes gramineus* **(Gmelin)***   VESP
Common summer resident north to southern Georgian Bay and Ottawa; rare north to Sioux Lookout and Timmins; formerly, at least, occasional to Moosonee. Occasional, rare in winter in the south (north to London and

Kingston). Common migrant. Principal dates in province: mid April to late October. Egg dates: 23 April to 3 August.
> *P. g. gramineus* breeds across the province west to Lake Nipigon, intergrading with the following subspecies to the Manitoba border.
>
> *P. g. confinis* intergrades with *gramineus* west of Lake Superior.

Lark Sparrow, *Chondestes grammacus* (Say)*   LASP
Occasional, rare summer resident in the south (north to Bruce County and Durham RM), also Sudbury; summer reports, north to Atikokan, Thunder Bay, and Moosonee. Vagrant in winter (resident in Lennox and Addington County; Wormington, 1987). Rare migrant. Principal dates in province: late April to late September. Egg dates: 15 May to 1 July.
> *C. g. grammacus.*

Lark Bunting, *Calamospiza melanocorys* Stejneger   LARB
Occasional, rare straggler, mainly in the south, but also north to Thunder Bay and North Point (James Bay). First reported in 1925 (Snyder, 1926), at least six reports in the past decade. Dates in province: all months but November; mainly spring and autumn.

Savannah Sparrow, *Passerculus sandwichensis* (Gmelin)*   SAVS
Abundant summer resident throughout the province. Occasional, rare winter resident in the extreme south (north to Peterborough and Kingston). Common migrant. Principal dates in province: mid April to late October. Egg dates: 27 April to 9 August.
> *P. s. mediogriseus* breeds in southern Ontario, intergrading with the following subspecies in central Ontario (Sault Ste. Marie, Wawa, Sudbury, and Biscotasing).
>
> *P. s. oblitus* breeds across northern Ontario, intergrading with the former subspecies in central Ontario.

[Baird's Sparrow, *Ammodramus bairdii* (Audubon)   BAIS
Five sight records since 1947, none of which has been accepted by OBRC.]

Grasshopper Sparrow, *Ammodramus savannarum* (Gmelin)*   GRSP
Uncommon local summer resident south of the Canadian Shield; rare north to Sudbury, and in western Rainy River District. Vagrant in winter in the south (Port Credit; Speirs, 1960b). Rare migrant. Dates in province: (early April) early May to mid September (mid November). Egg dates: 4 May to 12 August.
> *A. s. pratensis.*

Henslow's Sparrow, *Ammodramus henslowii* **(Audubon)***   **HESP**
Rare (to locally uncommon) summer resident in the south (north to Bruce County and Kingston). Rare migrant. Dates in province: (late March) early May to mid September (late December). Egg dates: 2 June to 14 August.
    A. h. henslowii.

Le Conte's Sparrow, *Ammodramus leconteii* **(Audubon)***   **LCSP**
Rare to locally uncommon summer resident at widely scattered locations in the north (from Thunder Bay and Timmins to the north coast); now virtually absent from the south, but reported from Dundas and Wellington counties and near Sudbury in recent years. Rare migrant. Dates in province: (early April) early May to early October (mid November). Egg dates: 3 to 16 June.

Sharp-tailed Sparrow, *Ammodramus caudacutus* **(Gmelin)***   **STSP**
Uncommon summer resident along the north coast (most numerous along the lower James Bay coast); summer records from western Rainy River District. Rare migrant. Dates in province: (late April) mid May to mid October (late December). Egg dates: 25 June to 16 July.
    A. c. altera breeds along the north coast.
    A. c. nelsoni occasional, rare migrant in the west and south; possible rare breeder in western Rainy River District.

Fox Sparrow, *Passerella iliaca* **(Merrem)***   **FOSP**
Uncommon summer resident in the north, south to Pickle Lake and Moosonee. Occasional, rare winter resident in the south (north to Bruce and Renfrew counties), also Kenora. Uncommon (to locally common) migrant. Principal dates in province: early April to late October. Egg dates: 2 June to 17 July.
    P. i. iliaca.

Song Sparrow, *Melospiza melodia* **(Wilson)***   **SOSP**
Abundant summer resident in the south, common in the north to the north shore of Lake Superior; becoming rare in the boreal forest and north to the Hudson Bay coast. Rare winter resident in the south (north to Manitoulin Island and Ottawa); occasional, rare in the north (Thunder Bay). Common migrant. Principal dates in province: mid March to mid November. Egg dates: 17 April to 3 September.
    M. m. euphonia breeds in the agricultural south, southwest from the Bruce Peninsula, Lake Simcoe, and Prince Edward County.
    M. m. melodia breeds in the south, west to Georgian Bay and Kingston, intergrading with *euphonia* south to Hamilton, and with *juddi* north of Algonquin Park.
    M. m. juddi breeds in the north, south to about Wawa and Lake Abitibi, then intergrades with *melodia* south to Algonquin Park.

Lincoln's Sparrow, *Melospiza lincolnii* **(Audubon)***   **LISP**
Common summer resident in the north, south to Thunder Bay and Kirkland Lake; becoming an occasional, rare and local summer resident in the south, mainly on the Canadian Shield, but south to Luther Marsh and Niagara RM. Occasional, rare in the south in winter (north to Peterborough). Uncommon migrant. Principal dates in province: mid May to early October. Egg dates: 16 May to 2 August.
   M. l. lincolnii.

Swamp Sparrow, *Melospiza georgiana* **(Latham)***   **SWSP**
Common (to abundant) summer resident throughout the province. Rare winter resident in the extreme south (north to Barrie and Kingston). Common migrant. Principal dates in province: early April to late December. Egg dates: 24 April to 25 July.
   M. g. ericrypta breeds in the north, south to about central Sudbury District where intergradation with *georgiana* begins.
   M. g. georgiana breeds in the south, intergrading with the former subspecies in southern Algoma and Sudbury districts.

White-throated Sparrow, *Zonotrichia albicollis* **(Gmelin)***   **WTSP**
Abundant summer resident throughout most of the province, but rare in the agricultural south, and almost completely absent from the Deciduous Forest Region. Rare winter resident in the south; occasional, rare in the north to Atikokan, Thunder Bay, and Moosonee. Abundant migrant. Principal dates in province: mid April to late December. Egg dates: 18 May to 8 August.

Golden-crowned Sparrow, *Zonotrichia atricapilla* **(Gmelin)**   **GCSP**
Occasional, rare straggler in the south, also Thunder Bay. First confirmed record in 1982, (ROM PR 1330–1341) Northumberland County, 3 to 15 January (Harris, 1983). Eight subsequent reports. Dates in province: early to mid January; mid April to early May; early October to early November.

White-crowned Sparrow, *Zonotrichia leucophrys* **(Forster)***   **WCSP**
Uncommon summer resident in the northern Hudson Bay Lowland, south to Big Trout Lake, and along the James Bay coast to Moosonee. Rare winter resident in the south (north to Sudbury). Common migrant. Principal dates in province: early May to late October. Egg dates: 17 June to 24 July.
   Z. l. leucophrys breeds in the extreme east, intergrading extensively with *gambelii* across the north.
   Z. l. gambelii breeds in the extreme west and intergrades extensively across the north.

Harris' Sparrow, *Zonotrichia querula* **(Nuttall)*** HASP
Rare summer resident in the extreme northwest (Fort Severn). Rare winter resident in the south (north to Peterborough), and west (Atikokan and Thunder Bay). Rare migrant in the east; uncommon in the west. Principal dates in province: late April to late October. Egg date: 5 July.

Dark-eyed Junco, *Junco hyemalis* **(Linnaeus)*** DEJU
Common summer resident in the north; uncommon in the south, mainly in the Algonquin Highlands, but south to Barrie and Kingston; an isolated few in Haldimand-Norfolk RM. Common winter resident in the south; occasional, rare north to Thunder Bay and Moosonee. Common (to locally abundant) migrant. Egg dates: 18 April to 26 July.
  J. h. hyemalis breeds.
  J. h. cismontanus occasional, rare straggler, autumn to spring.
  J. h. montanus occasional, rare straggler, autumn to spring.
  J. h. shufeldti occasional, rare straggler, autumn to spring.

Lapland Longspur, *Calcarius lapponicus* **(Linnaeus)*** LALO
Common summer resident in the Tundra Region. Uncommon (to locally common) winter resident in the south (north to Kingston). Uncommon (to locally abundant) migrant. Egg dates: 17 June to 21 July.
  C. l. lapponicus.

Smith's Longspur, *Calcarius pictus* **(Swainson)*** SMLO
Common summer resident in the Tundra Region. Migrates west from Ontario; occasional, rare (to locally uncommon) south of the tundra, principally in the Lake Superior area. Dates in province: (mid April) late May to mid September (early November). Egg dates: 22 June to 14 July.

Chestnut-collared Longspur, *Calcarius ornatus* **(Townsend)** CCLO
Vagrant; one photographic record: (ROM PR 1077–1082) Sudbury, 17 April 1978 (Goodwin, 1979b). One sight record: Kingston, 2 May 1972 (Weir, Quilliam, and Norman, 1972).

Snow Bunting, *Plectrophenax nivalis* **(Linnaeus)[*]** SNBU
Occasional, rare summer resident along the north coast (Peck, 1972; Cadman, Eagles, and Helleiner, 1987). Common (to abundant) winter resident in the south; rare in the west (Thunder Bay). Common (to locally abundant) migrant. Principal dates in province: mid October to early May.
  P. n. nivalis.

Bobolink, *Dolichonyx oryzivorus* **(Linnaeus)*** BOBO
Common summer resident in agricultural areas of the south, rare in the north to Dryden and Kapuskasing; vagrant to Big Trout Lake and Moo-

sonee. Common (to locally abundant) migrant. Dates in province: (early April) early May to mid September (mid December). Egg dates: 19 May to 16 July.

Red-winged Blackbird, *Agelaius phoeniceus* **(Linnaeus)***   **RWBL**
Abundant summer resident in the south; locally common in the north to Sandy Lake and Moosonee; rare north to Hudson Bay. Uncommon winter resident in the south (north to Toronto); occasional, rare (to uncommon) north to Thunder Bay and Moosonee. Abundant migrant. Principal dates in province: early March to late December. Egg dates: 5 April to 3 August.
- *A. p. arctolegus* breeds across the province east to Nipigon and Moosonee, intergrading to the southeast with *phoeniceus*.
- *A. p. phoeniceus* breeds in the south, and in the north to Chapleau and Cochrane, intergrading to the northwest with *arctolegus*.

Eastern Meadowlark, *Sturnella magna* **(Linnaeus)***   **EAME**
Common summer resident in agricultural areas of the south; rare in the north to Thunder Bay and western Rainy River District, and in the extreme east to agricultural areas of the Clay Belt (to northern Timiskaming District). Rare winter resident in the south (north to Kingston); vagrant in the north (Dorion). Common migrant. Principal dates in province: mid March to mid November. Egg dates: 2 May to 3 August.
- *S. m. magna*.

Western Meadowlark, *Sturnella neglecta* **Audubon***   **WEME**
Common summer resident in western Rainy River District; rare in the Thunder Bay area; rare (to locally uncommon) in the south in agricultural areas, mainly west of Toronto. Rare migrant; wandering north to Sandy Lake and Moosonee. Dates in province: (mid March) early April to mid October (late December). Egg dates: 16 May to 3 July.
- *S. n. neglecta*.

Yellow-headed Blackbird,   **YHBL**
Xanthocephalus xanthocephalus **(Bonaparte)***
Locally uncommon summer resident in the west at Rainy River and Atikokan; rare at Thunder Bay; locally uncommon in the south (Essex and Kent counties); possibly in Haldimand-Norfolk RM. Summer records from Red Lake, Longlac, and Simcoe County. First breeding reported in 1961 (Baillie, 1961); slowly expanding in Ontario; sight records north to Moosonee in spring, summer, and autumn. Occasional, rare in winter in the south. Rare migrant. Principal dates in province: early May to late October. Egg dates: 30 May to 1 July.

Rusty Blackbird, *Euphagus carolinus* (Müller)*   RUBL
Uncommon summer resident across the north, and in the Algonquin Highlands in the south; occasional, rare south to the Bruce Peninsula, Haliburton District, and Ottawa. Rare (to common) winter resident in the south (north to Manitoulin Island and Ottawa); occasional, rare north to Thunder Bay, Wawa, and Moosonee. Common migrant. Principal dates in province: late March to mid November. Egg dates: 4 May to 22 June.
   E. c. carolinus.

Brewer's Blackbird, *Euphagus cyanocephalus* (Wagler)*   BRBL
Locally uncommon summer resident, Kenora to Thunder Bay, Sault Ste. Marie to North Bay, Manitoulin Island, and the Bruce Peninsula; rare, north to Red Lake, Geraldton, and Timiskaming District, south to Point Pelee, east to Prince Edward county. Occasional, rare in winter in the south (north to Manitoulin Island); also in the north at Thunder Bay, Schreiber, and Matachewan. Rare (to locally uncommon) migrant. Principal dates in province: early April to mid November. Egg dates: 6 May to 26 June.

Great-tailed Grackle, *Quiscalus mexicanus* (Gmelin)   GTGR
Vagrant; two photographic records: (ROM PR 2033–2045) Atikokan, 7 to 25 October 1987 (Elder, 1988); Port Rowan, 19 November 1988 to 6 January 1989 (Weir, 1989a, 1989c).

Common Grackle, *Quiscalus quiscula* (Linnaeus)*   COGR
Abundant summer resident in the south, locally common in the north to Big Trout Lake and Fort Albany; vagrant to Winisk (Schueler, Baldwin, and Rising, 1974). Rare (to locally common) winter resident in the south; occasional, rare (to uncommon) in the north to Thunder Bay and Moosonee. Abundant migrant. Principal dates in province: early March to late December. Egg dates: 4 April to 12 July.
   Q. q. versicolor.

Brown-headed Cowbird, *Molothrus ater* (Boddaert)*   BHCO
Common summer resident across the province, north to Kenora, Lake Nipigon, and Cochrane; rare to Pickle Lake and Moosonee. Rare to uncommon winter resident in the south (north to Ottawa); occasional, rare in the north (Thunder Bay). Common (to locally abundant) migrant. Principal dates in province: mid March to late December. Egg dates: 17 April to 5 August.
   M. a. ater breeds across the province, west to Lake Nipigon, intergrading with *artemisiae* in the west.
   M. a. artemisiae intergrades with *ater* in the west, east to Lake Nipigon; vagrant at James Bay (Godfrey, 1951; Manning, 1952).

Orchard Oriole, *Icterus spurius* **(Linnaeus)*** OROR
Uncommon local summer resident in the Deciduous Forest Region; rare north to Georgian Bay and Kingston. Uncommon migrant; vagrant in the north (Terrace Bay). Dates in province: (mid April) mid May to late August (late September). Egg dates: 24 May to 10 July.

Northern Oriole, *Icterus galbula* **(Linnaeus)*** NOOR
Common summer resident north to Georgian Bay and Ottawa, uncommon north to Kenora, Thunder Bay, and Sudbury; occasional, rare to Red Lake, Hearst, and Timiskaming District. Occasional, rare winter resident in the south (north to Sault Ste. Marie and Kingston). Common migrant; wanders north to Moosonee. Principal dates in province: early May to early September. Egg dates: 18 May to 24 June.
  I. g. galbula breeds.
  I. g. bullockii vagrant at Thunder Bay (1977) and in the south (north to Bruce County and Peterborough). First reported in 1975 (Goodwin, 1976c); at least four reports in the past decade.

Scott's Oriole, *Icterus parisorum* **Bonaparte** SCOR
Vagrant; one photographic record: (ROM PR 472–474) Silver Islet, Thunder Bay District, 9 November 1975 (Denis, 1976).

Brambling, *Fringilla montifringilla* **Linnaeus** BRAM
Vagrant; two photographic records: (ROM PR 1457–1464, 1632–1637) Atikokan, 23 to 26 October 1983 (James, 1984b); (ROM PR 2048) Brampton, Peel RM, 12 to 18 November 1980 (Coady and Wormington, 1989).

Rosy Finch, *Leucosticte arctoa* **(Pallas)** ROFI
Occasional, rare straggler in winter, usually in the west; three records from Thunder Bay (first in 1963; Baillie, 1964), two from Dryden, and one from Manitoulin Island. Dates in province: early November to late March.
  L. a. littoralis two photographic records.
  L. a. tephrocotis one photographic record.

Pine Grosbeak, *Pinicola enucleator* **(Linnaeus)*** PIGR
Rare (to uncommon) summer resident across northern Ontario. Common winter erratic in the southern half of northern Ontario, becoming rare farther north; uncommon winter erratic in the south (to Lake Erie). Egg dates: unknown.
  P. e. leucura.

Purple Finch, *Carpodacus purpureus* **(Gmelin)*** PUFI
Rare summer resident south of the latitude of Toronto, becoming common north to Sandy Lake and Moosonee; occasional, rare north to Fort Severn and Sutton Lake. Rare to common winter erratic in the south (north to

Manitoulin Island); also occasional, rare (to uncommon) to Thunder Bay in the north. Uncommon to common (rarely locally abundant) migrant. Egg dates: 17 May to 5 August.

C. p. purpureus.

House Finch, *Carpodacus mexicanus* **(Müller)***   **HOFI**
Common (to locally abundant) permanent resident near Lakes Erie and Ontario, summering rarely north to Georgian Bay and Renfrew County. A rapidly expanding species first reported in the province in 1970 (Weir, 1970); breeding verified for the first time in 1978 (James, 1978). Vagrant in the north to Marathon and Timiskaming District. Uncommon (to locally common) migrant. Egg dates: 21 March to 3 August.

C. m. frontalis.

Red Crossbill, *Loxia curvirostra* **Linnaeus***   **RECR**
Uncommon, permanent erratic resident across the province, north to Sandy Lake and Cochrane, south to Algonquin Park. Occasional, rare in summer north to Fort Severn and Sutton Lake, south to Haldimand-Norfolk RM. Rare to common erratic south to Lake Erie, autumn to spring. Egg dates: 4 to 18 April.

L. c. minor breeds.
L. c. bendirei rare winter erratic.
L. c. pusilla rare winter erratic.
L. c. benti rare winter erratic.
L. c. sitkensis rare winter erratic.

White-winged Crossbill, *Loxia leucoptera* **Gmelin***   **WWCR**
Uncommon to common, permanent erratic resident across the province, south to the Algonquin Highlands; occasional farther south (to Oxford County and Lake Ontario) in summer. Rare to locally abundant erratic south to Lake Erie, autumn to spring. Egg date: 19 August.

L. l. leucoptera.

Common Redpoll, *Carduelis flammea* **(Linnaeus)***   **CORE**
Common summer resident along the Hudson Bay coast; occasional, rare south to Attawapiskat Lake and Moosonee; summer sight records of non-breeding (?) birds from Kenora, Thunder Bay, and Smoky Falls. Uncommon to common erratic throughout all but the far north of the province, late autumn to early spring. Common to locally abundant migrant in the north. Egg dates: 16 June to 22 July.

C. f. flammea breeds, migrant and winter erratic.
C. f. rostrata erratic, autumn to spring.

Hoary Redpoll, *Carduelis hornemanni* **(Holböll)[*]**   HORE
Rare (and occasional?) summer resident on the north coast (Cadman, Eagles, and Helleiner, 1987). Rare (to uncommon) erratic, late autumn to early spring, south to Lake Ontario, occasional to Lake Erie. Principal dates in province: late October to mid April.
   C. h. hornemanni erratic, autumn to spring (breeding?).
   C. h. exilipes erratic, autumn to spring (breeding?).

Pine Siskin, *Carduelis pinus* **(Wilson)***   PISI
Common summer resident across the province, north to Sandy Lake and Moosonee, south to Lake Simcoe and Ottawa; occasional, rare north to Fort Severn and Attawapiskat, south to Lake Erie. Common to locally abundant winter erratic in the south; becoming occasional, north to Sioux Lookout and Cochrane. Locally common (to abundant) migrant. Egg dates: (?) March to 23 July.
   C. p. pinus.

Lesser Goldfinch, *Carduelis psaltria* **(Say)**   LEGO
Vagrant; one sight record: Toronto, 10 August 1982 (James, 1983); it was accepted by OBRC, but I have some doubts that it was a natural occurrence.

American Goldfinch, *Carduelis tristis* **(Linnaeus)***   AMGO
Common summer resident north to Sioux Lookout and Cochrane; occasional, rare to southern James Bay; vagrant at Attawapiskat. Common winter resident in the south (north to Georgian Bay and Ottawa); becoming occasional, north to Kenora and Cochrane. Common (to locally abundant) migrant. Egg dates: 13 June to 24 September.
   C. t. tristis breeds across the province, west to Kenora, intergrading with *pallida* west of Lake Superior.
   C. t. pallida breeds in the extreme west.

Evening Grosbeak, *Coccothraustes vespertinus* **(Cooper)***   EVGR
Common summer resident across the province, north to Pickle Lake and Moosonee, south to Simcoe County and Brockville; occasional, rare to Haldimand-Norfolk RM. Common winter resident in the south; uncommon in the northern summer range, north to about the latitude of northern Lake Superior. Common (to locally abundant) migrant. Egg dates: 13 June to 4 July.
   C. v. vespertinus.

House Sparrow, *Passer domesticus* **(Linnaeus)***   HOSP
Common to locally abundant permanent resident, north to Georgian Bay and Ottawa; becoming increasingly uncommon and local, north to Red Lake and Moosonee; vagrant to the north coast. Egg dates: 4 April to 14 August. Introduced to Ontario about 1870.
   P. d. domesticus.

ACKNOWLEDGEMENTS

The previous checklist, prepared by Ross James, Peter McLaren, and Jon Barlow (James, McLaren, and Barlow, 1976), drawing on an extensive background of information compiled by Lester Snyder and Jim Baillie, formed the framework on which the present checklist was based. However, in the past dozen years, the contributions of hundreds of individuals to the Royal Ontario Museum, to seasonal summaries in *American Birds,* to the Ontario Breeding Bird Atlas (Cadman, Eagles, and Helleiner, 1987), and to the Ontario Bird Records Committee, have made changes to the information of almost every species. Without the input of these many unnamed individuals, this rewriting would not have been meaningful.

I am grateful to Nick Escott, George Peck, Al Sandilands, Don Sutherland, and Ron Weir for comments on the manuscript. Seasonal summaries in American birds prepared by Ron Weir, and reports forwarded to the museum by him, have been most helpful. Henri Ouellet also provided helpful comments, particularly with respect to subspecies, and made possible access to specimens and records in the Canadian Museum of Nature. Earl Godfrey provided information and comments about species status and distribution, and *The Birds of Canada* (Godfrey, 1986) formed a basic reference for consideration of subspecific distributions. The preparation of the manuscript was aided by the librarians and members of the photography department of the ROM. Marg Goldsmith assisted in manuscript preparation.

Publication was supported in part by a generous grant from the Ontario Ministry of Natural Resources.

APPENDICES

Appendix I: Probable Escapees

Published reports of the following species, based on specimens, photographs, or sight records, are known or believed to represent occurrences of escaped or released captives, rather than natural events.

Scarlet Ibis, *Eudocimus ruber* **(Linnaeus)**
Point Pelee sight record, September 1937 (Stirrett, 1941).

Whooper Swan, *Cygnus cygnus* **(Linnaeus)**
Western Lake Ontario photographic and sight records, winters of 1978–1979, 1980, and 1981 (Goodwin, 1979a, 1980a, 1981).

Bean Goose, *Anser fabalis* **(Latham)**
Port Colborne sight record, April 1933 (Beardslee and Mitchell, 1965).

Ruddy Shelduck, *Tadorna ferruginea* **(Pallas)**
Long Point sight record, June 1988 (Weir, 1988c).

White-cheeked Pintail, *Anas bahamensis* **Linnaeus**
Amherstview sight record, July 1974 (Weir, 1974).

Ringed Turtle-Dove, *Streptopelia risoria* **(Linnaeus)**
Numerous reports from Thunder Bay southward, even of nesting pairs. This species is not established in the wild.

Black-hooded Parakeet, *Nandayus nenday* **(Vieillot)**
Windsor sight record, autumn 1971 (Goodwin, 1973a).

Scrub Jay, *Aphelocoma coerulescens* **(Bosc)**
Photographic records: Halton County, August 1985 to March 1986; Northumberland County, May 1986 (Wormington, 1987).

Chihuahuan Raven, *Corvus cryptoleucus* **Couch**
Long Point photographic record, May 1976 (Goodwin, 1976b).

Blue Tit, *Parus caeruleus* **Linnaeus**
Gravenhurst photograph (ROM PR 368), October 1973 to April 1974 (Goodwin, 1974a).

Pyrrhuloxia, *Cardinalis sinuatus* **Bonaparte**
Port Credit sight record, November 1959 (Speirs, 1960a).

Serin, *Serinus serinus* **(Linnaeus)**
Pickering sight record, December 1985 to March 1986 (Wormington, 1987).

Eurasian Siskin, *Carduelis spinus* **(Linnaeus)**
Toronto photographic record, February 1988 (Weir, 1988a).

European Goldfinch, *Carduelis carduelis* **(Linnaeus)**
Toronto specimen, May 1887 (Fleming, 1907), and numerous sight records to the present.

Linnet, *Carduelis cannabina* **(Linnaeus)**
Toronto sight record, January 1890 (Fleming, 1907).

Appendix II: Comments on Subspecific Variation

Comments follow on the subspecific affinities of Ontario birds, where this checklist is at variance with the fifth edition of the AOU checklist (AOU, 1957) or the previous Ontario checklist (James, McLaren, and Barlow, 1976), or where subsequent studies indicate changes should be considered.

Northern Fulmar, *Fulmarus glacialis*
A single specimen (ROM 154524), collected 15 January 1989 at Presqu'ile Provincial Park, is the only indication of *F. g. glacialis* in Ontario (culmen 40.7 mm, bill width at base 21.0 mm).

Fulvous Whistling-Duck, *Dendrocygna bicolor*
Palmer (1976) indicates that no subspecies are recognized by recent authorities.

Greater Snow Goose, *Chen caerulescens atlantica*
Definitive evidence of the occurrence of this form in Ontario is currently lacking, in my estimation. Only one specimen labelled *atlantica* is available, a large-billed bird in ROM collections: it is within the size range of *C. c. caerulescens* given by Palmer (1976), and was taken in summer where *C. c. caerulescens* is expected.

Canada Goose, *Branta canadensis*
As indicated by Palmer (1976), the nomenclature and distinctiveness of races are subject to disagreement and further revision. Racial identities have been broken down because of local extinctions, releases of questionable origin, escaped captives, and altered migration routes; however, as the designations provided by Palmer (1976) allow reasonable separation on the basis of measurements and plumages, I have used them as the basis for my assessment and nomenclature.

Specimens labelled *B. c. parvipes* in ROM collections all have measurements that fall within Palmer's range for *B. c. hutchinsii;* hence I am not aware of specimens of *parvipes* from Ontario, although Bellrose (1976) indicates that birds of this form migrate through the province.

American Black Duck, *Anas rubripes*
In addition to the fact that these birds breed extensively with the Mallard, recent biochemical studies have indicated that these two forms are virtually indistinguishable genetically. Some now consider that the Black Duck should probably be consid-

ered only a melanistic morph of the Mallard (Ankney et al., 1986; Avise, Ankney, and Nelson, 1990).

Blue-winged Teal, *Anas discors*
Birds from Ontario (James Bay) labelled *A. d. orphna* are indistinguishable from any other Ontario Blue-winged Teal.

Common Eider, *Somateria mollissima*
The relative frequency of *S. m. sedentaria* and *S. m. dresseri* in southern Ontario is uncertain, as very few winter specimens have ever been collected. The winter specimen from Lake Simcoe is *sedentaria* in my opinion. I have not examined a Niagara specimen in the Buffalo Museum assigned to *sedentaria* by Snyder (unpubl. ms.).

Bald Eagle, *Haliaeetus leucocephalus*
Robbins (1960) indicated that banded eagles raised in Florida, *H. l. leucocephalus*, had been recovered in Ontario. Two immature specimens in the ROM conform to measurements of *leucocephalus* as given by Friedmann (1950).

Red-tailed Hawk, *Buteo jamaicensis*
There is no evidence that *B. j. calurus* breeds in Ontario. All adults from the breeding season (except *kriderii*) conform to *borealis*, not *calurus*. There is a nestling in ROM collections labelled *calurus*, but it is very similar to a nestling *kriderii* that was collected with an adult to confirm the identity.

Merlin, *Falco columbarius*
There are a number of pale-coloured migrants in ROM collections conforming to *F. c. bendirei*. These then cast doubt on the racial identity of the *richardsonii* sight record of Wormington (1984b).

Peregrine Falcon, *Falco peregrinus*
Birds from Ontario, typical of those described as *F. p. tundrius* (White, 1968), are readily distinguished from *F. p. anatum* and merit subspecific definition in my opinion. There are also several specimens of plumage that is intermediate between typical *tundrius* or *anatum* formerly taken as migrants in Ontario. The zone of intergradation was unlikely to have been in Ontario.

Gray Partridge, *Perdix perdix*
I have insufficient comparative material to attempt to assign subspecific identities to Ontario birds, but there are substantial differences in some from *P. p. perdix* of England. A mixture of varieties is probable as birds from several European countries were introduced (Snyder, unpubl. ms.).

Ring-necked Pheasant, *Phasianus colchicus*
Introductions were from a variety of places and no identifiable subspecies can be recognized.

Spruce Grouse, *Dendragapus canadensis*
The analysis by Rand (1948) presents a good assessment of variation in Ontario. There is a wide belt of intergradation between *D. c. canadensis* and *D. c. canace*

across the province and, with much individual variation, the two forms are distinctive only at the extreme ends of their ranges. Most birds in southern Ontario conform to *canace,* but individuals of this type may be found north to Lac Seul and southern James Bay. Birds more similar to *canadensis* are found south to the Georgian Bay area.

Willow Ptarmigan, *Lagopus lagopus*
The single bird from Whitby previously listed as *L. l. ungavus* cannot be placed in any subspecies on the basis of bill measurements.

Ruffed Grouse, *Bonasa umbellus*
Extensive intergradation makes areas of separation somewhat arbitrary. Typically, birds from northern Ontario considered by Todd (1940, 1947) to be a separate form, *obscura,* are readily separable from *togata.* The forms *monticola* and *togata* are more difficult to separate. The *obscura* form is, if anything, doubtfully separated from *umbelloides.* Although typical forms of each can be distinguished, they intergrade gradually from east to west. Birds labelled *umbelloides* from western Ontario are closer to *obscura* in form than to typical *umbelloides.* A single pale gray bird (*incana*) from western Ontario is notably distinct from any of the other Ontario birds.

Sharp-tailed Grouse, *Tympanuchus phasianellus*
Birds typical of *T. p. campestris* obviously occur in the Kenora to Rainy River area and probably breed there, but the few specimens from the breeding season are more typical of *T. p. phasianellus.* These two forms apparently interbreed in the area.

Birds typical of *campestris* occur and probably breed in southern Algoma and Manitoulin districts, but material is all collected in autumn. A single bird labelled *T. p. jamesi* in the ROM is a mislabelled *campestris,* in my opinion.

Sandhill Crane, *Grus canadensis*
Although measurements given by Walkinshaw (1965) seem to provide reasonably good separation of three subspecies that affect Ontario, considerable doubt has been expressed as to whether morphological variation really reflects geographic distribution (Tacha, Vohs, and Warde, 1985). This species has only recently moved (back) into Ontario in numbers (Lumsden, 1971; Tebbel, 1981), and there are few specimens on which to base any subspecific determination, even if the named forms prove to be distinguishable.

Piping Plover, *Charadrius melodus*
As the extent of belting on the breast is of doubtful significance in delineating subspecies (Wilcox, 1959), as both belted and unbelted birds have been found in Ontario, and as there is some mixing between populations in eastern and central North America (Haig and Oring, 1988), I see no way to distinguish racial differences based on such a character.

Ruddy Turnstone, *Arenaria interpres*
Birds from Ontario show considerable variation, and could even represent intergradation between *A. i. interpres* and *A. i. morinella* if not pure *interpres.* If these subspecies can be separated, I cannot be certain what the Ontario specimens represent without additional material for comparison.

Short-billed Dowitcher, *Limnodromus griseus*
Few specimens have ever been taken in Ontario in summer, indicating that there is, at best, a thin scattering of nesting birds. Ontario lies between the main breeding ranges of two subspecies; birds from Ontario can appear similar to either type, or are of intermediate plumage (also indicated by Pitelka, 1950), suggesting that there is a mixture of both here.

Thayer's Gull, *Larus thayeri*
I prefer to follow Godfrey (1986) in considering this form a subspecies of Iceland Gull, *L. glaucoides*. I have not personally undertaken a thorough review, but specimens in the NMC and the ROM support this conclusion, as do the observations of Gaston and Decker (1985), and Snell (1989).

Lesser Black-backed Gull, *Larus fuscus*
All specimens, photographs, and written reports available indicate that birds are the slate-gray-backed form, *graellsii,* with one or two possible exceptions (Weir, 1987b; and an unpublished description in the ROM).

Black Guillemot, *Cepphus grylle*
I have inadequate material to review subspecific variation. Salomonson (1944) and Storer (1952) have identified the largely sedentary Hudson/James Bay population as belonging to *ultimus*. It is possible that some of the winter occurrences in the south are *atlantis*.

Band-tailed Pigeon, *Columba fasciata*
Although ROM collections include no *C. f. fasciata* for comparison, the Ontario specimen is darker overall than other *monilis* from British Columbia.

Mourning Dove, *Zenaida macroura*
Birds typical of and intermediate between *marginella* and *carolinensis* occur throughout Ontario in summer and during migration seasons. Most of the few available from west of Lake Superior resemble *marginella*. If these two forms are maintained, then Ontario must fall in a zone of overlap and intergradation, as indicated by Aldrich, Duvall, and Gies (1958).

Great Horned Owl, *Bubo virginianus*
Birds from most of northern Ontario are distinctively different from *B. v. virginianus,* as described by Snyder (1961), and I accept his designation, *scalariventris*. The much more rufous-coloured *virginianus* of southern Ontario is easily distinguished, although in the northern parts of southern Ontario and along the southern fringes of northern Ontario there is some rufous influence in *scalariventris*. Toward the western part of the province, birds appear to be much lighter and approach the prairie birds, *subarcticus,* in colour (*subarcticus* rather than *wapacuthu* as per Manning 1952, pp. 63–64). There are no birds in ROM collections resembling *subarcticus* in summer from northern Ontario, north of the range of *scalariventris,* to indicate that *subarcticus* breeds across the far north, but it may be the usual form in the northwest.

A single bird from Ontario in winter suggested to be *occidentalis* (James, McLaren, and Barlow, 1976) is referable to *subarcticus*-like birds that occur in the western part of the province. It is slightly more rufous than typical *subarcticus,* as are *scalariven-*

tris and *subarcticus* from the southern parts of western Ontario. However, I agree with Godfrey (1986) that prairie birds in Canada are very *subarcticus*-like, and noticeably paler than *occidentalis*. Southern prairie birds may be more rufous than *subarcticus* from farther north, as shows in western Ontario summer birds.

Red-headed Woodpecker, *Melanerpes erythrocephalus*
I see no reason to place birds in western Ontario into a separate subspecies. Although the sample size is small (four males and three females), the wing measurement of only one bird exceeds those of *M. e. erythrocephalus* as given by Brodkorb (1935), while wing measurements of three others are below the lower limits given for *M. e. caurinus*.

Downy Woodpecker, *Picoides pubescens*
I see no significant and consistent differences between the Downy Woodpeckers from any part of the province, and all should be referred to one subspecies. Measurements and colour characteristics overlap throughout the province. Ouellet (1977) also found only *medianus* in Ontario, with birds from north of the latitude of Lake Nipigon more intermediate in characteristics between *medianus* and *nelsoni*.

Hairy Woodpecker, *Picoides villosus*
I see little significant or consistent difference between northern and southern birds. Northern birds may average slightly larger, but shorter-winged birds can be found north to James Bay and longer-winged birds south to Lake Ontario. Measurements for any specimen scarcely reach 125 mm, or well below the size given by Ridgway (1914) for the *septentrionalis* race, and even below the arbitrary size given by Snyder (1953). No consistent colour differences were noted either. However, a more extensive study by Ouellet (1977) indicates that two subspecies, with centres of distribution either side of Ontario, form a broad band of intergradation across the central part of the province.

Northern Flicker, *Colaptes auratus*
While there may be slight average differences in wing measurement between birds of southern and northern Ontario, as noted by Godfrey (1986), these are insufficient to distinguish two "yellow-shafted" subspecies clearly.

While several "yellow-shafted" birds with a few reddish primary shafts have been taken in Ontario, the only evidence for a "red-shafted" form is the single bird photographed at Thunder Bay, 20 November 1988. From the photographs the subspecies is uncertain, but appears to be *collaris*.

Horned Lark, *Eremophila alpestris*
Birds from northern Ontario may be similar to either *hoyti*, paler yellowish about the head and with a less vinaceous colour, or to *alpestris*, with strong yellows and darker backs. Most, however, are intermediate with rather paler yellows and more vinaceous backs.

The birds from western Rainy River, considered by Snyder (1938) to be *enthymia*, are much darker than typical *enthymia* from the central prairies, are darker than birds from Manitoba, and are scarcely lighter than birds from southern Ontario. These western birds may be somewhat intermediate, but are more like *praticola* than *enthymia*, in my estimation.

Cliff Swallow, *Hirundo pyrrhonota*
The supposed *hypopolia* from Fort Severn has a pale rump, but otherwise conforms to the nominate form as found throughout Ontario.

Carolina Chickadee, *Parus carolinensis*
Several recent studies clearly confirm interbreeding with Black-capped Chickadees (Robbins, Braun, and Tobey, 1986; Braun and Robbins, 1986; Mack et al., 1986).

Bewick's Wren, *Thryomanes bewickii*
Perhaps there is insufficient material, but I can see no substantive differences between the races *altus* and *bewickii* among birds so labelled in ROM collections.

House Wren, *Troglodytes aedon*
Although Godfrey (1986) suggests merging *baldwini* and *aedon,* there seems to me to be reason for maintaining them. Birds from Manitoulin Island are quite brown, but birds from the Sault Ste. Marie and southern Sudbury districts show definitely grayer plumages, as do some from Parry Sound District.

Veery, *Catharus fuscescens*
In eastern Ontario, south or north, birds tend to be less olivaceous than the western *salicicola,* but there is considerable variation, with birds from more northern localities being browner than southern birds. Although birds referable to *salicicola* have been taken at Sault Ste. Marie and on Manitoulin Island in the breeding season, these seem to be unusual. A single bird from Rossport on northern Lake Superior is slightly more olivaceous rather than richer brown. More material is needed from northern Ontario but, from available material, I believe *salicicola* is primarily western and *fuscescens* eastern, with a few *salicicola*-like birds penetrating at Sault Ste. Marie.

Gray-cheeked Thrush, *Catharus minimus*
Small birds referable to *bicknelli,* primarily on the basis of size, occur in southern Ontario during migration. I agree with Godfrey (1986) that the rest in Ontario should be *aliciae,* as the few Newfoundland birds in ROM collections are notably different (= *minimus*).

Brown Thrasher, *Toxostoma rufum*
If the race *longicauda* is upheld, there is insufficient material from western Ontario to indicate that birds there are any different from birds in the rest of the province, but they may be intermediate.

Northern Shrike, *Lanius excubitor*
Birds from Ontario are consistently darker than birds from much farther west in Canada. Birds from Manitoba are a shade lighter than in Ontario, but darker than far western birds. Ontario may be an area of transition from eastern to western forms, but birds here are more similar to *borealis.*

Yellow Warbler, *Dendroica petechia*
Birds in northern Ontario are rather uniformly darker, including those in western Rainy River District. A few lighter birds appear as far north as Attawapiskat. In

southern Ontario many birds as far south as Lake Erie are scarcely lighter than in the north, and even the lighter *aestiva*-like birds are not as pale as birds from the southern United States. Southern Ontario seems more of an area of intergradation between *aestiva* and *amnicola*.

American Redstart, *Setophaga ruticilla*
I see no reason for considering birds from the extreme south of the province as any different from the rest. Although I have continued to use *tricolora*, I do not have sufficient material to know whether it should be maintained.

Ovenbird, *Seiurus aurocapillus*
Variation in the colour of Ontario birds, especially immatures and fresh-plumaged adults, overlaps that of *furvior*, so that it is not possible to identify any single migrant as belonging to the Newfoundland race.

Northern Waterthrush, *Seiurus noveboracensis*
There is no indication of the eastern race *S. n. noveboracensis* occurring in Ontario among specimens in the ROM or the NMC. Birds from northern Ontario are frequently as yellow below as birds from Newfoundland, but all seem to be grayer above than is typical of Newfoundland birds.

MacGillivray's Warbler, *Oporornis tolmiei*
Specimen not examined.

Common Yellowthroat, *Geothlypis trichas*
Differences between eastern and western birds in Ontario are scarcely perceptible. There is a tendency for birds west of Lake Superior to be lighter on the back and very slightly more golden-yellow below, but they are not as light or as yellow as birds from farther west in Canada.

Rufous-sided Towhee, *Pipilo erythrophthalmus*
Heavily spotted birds of the western *arcticus/montanus* type have appeared in Ontario on a number of occasions, but none have been collected for positive identification. A careful description of one bird, watched closely for several days by H. H. Axtell, and compared to museum skins at the time, was of the *montanus* type. Others, especially in western Ontario, are more likely to have been of the *arcticus* subspecies.

Chipping Sparrow, *Spizella passerina*
Birds from west of the longitude of Lake Nipigon show intermediate and variable colour, but tend to be more often somewhat lighter, approaching the western *boreophila* form, as the Manitoba boundary is approached.

Vesper Sparrow, *Pooecetes gramineus*
Most birds in Ontario west of Lake Superior show variable colour, from typical *gramineus* to typical *confinis*. Very few are as pale as *confinis* from the central prairies. The whole of western Ontario seems to be a zone of intergradation, as suggested by Snyder (1953).

Savannah Sparrow, *Passerculus sandwichensis*
A few of the birds from Moosonee to Cape Henrietta Maria are somewhat darker and more like *labradorius,* but most are typical of *oblitus.* From Sault Ste. Marie and Sudbury north to about Wawa and Biscotasing, a mixture of *oblitus* and *mediogriseus* is found.

Grasshopper Sparrow, *Ammodramus savannarum*
Several migrants from southern Ontario appear much like the birds from Manitoba considered to be *perpallidus;* however, I find them darker than birds from farther south and west. There are no specimens and few positive sight records from western Ontario; however, if they occur there, they are probably intermediate between typical *pratensis* and *perpallidus,* as are the adjacent Manitoba birds.

Sharp-tailed Sparrow, *Ammodramus caudacutus*
One observer has apparently located a breeding colony in western Rainy River District (unpublished). The area may be occupied only occasionally; no birds have been collected, nor proof of breeding established, but they are probably of the *nelsoni* race.

Song Sparrow, *Melospiza melodia*
All birds from Lake Superior north are darker, like *juddi.* From Wawa and Lake Abitibi south to Algonquin Park there is a mixture of *juddi, melodia,* and intermediate birds. In southern Ontario to the north and east of Lake Simcoe and Prince Edward County, largely on the Canadian Shield, birds are *melodia*-like. In the agricultural areas south and west of the Shield, most birds are *euphonia*-like, although there are many intermediate and *melodia*-like birds south to Hamilton and Huron County.

White-crowned Sparrow, *Zonotrichia leucophrys*
Most of the northern Ontario birds do not show a distinctly cut-off supercilliary in front of the eye. There is usually a constriction of the supercilliary and various shades of gray in front of the eye. As far east as Hawley Lake, a clear white supercilliary extending to the bill is seen on some birds, as is typical of *gambelii.* Most of northern Ontario, then, is a zone of intergradation.

Dark-eyed Junco, *Junco hyemalis*
Although there are several reports of "Oregon" juncos in the Ontario literature, I can find no specimens typical of *J. h. oreganus.* Likewise, *mearnsi* has been reported as sighted in Ontario, but no specimens confirm this. Birds with pinky or rufescent colours, typical of birds of the western mountains, are found in Ontario from autumn to spring and correspond to one of *cismontanus, montanus,* or *shufeldti* groups.

Brown-headed Cowbird, *Molothrus ater*
Birds from the extreme western part of Ontario are intermediate in colour between typical prairie and typical eastern plumages.

Pine Grosbeak, *Pinicola enucleator*
Perhaps I have inadequate material, but I cannot substantiate the occurrence of *eschatosa* in Ontario, if it is a valid race. Among summer-collected grosbeaks there is a range of colour, with the darkest being as dark as Newfoundland birds. A male

from near Hudson Bay is not only as intensely red, but also as small. Do small dark winter birds, then, represent wanderers from the east, or just the normal complement of Ontario birds?

Red Crossbill, *Loxia curvirostra*

There are very few nesting or breeding records for Ontario, and birds from any of them were never collected for subspecific identification. While there is need for more extensive studies, summer-collected birds indicate that Ontario lies within the usual range of *minor,* and the other subspecies are ordinarily only irregular winter visitors. However, the presence of some juvenile birds labelled *minor,* with some summer-collected adults labelled *sitkensis* (both labelled by A. R. Phillips), suggests either that some wanderers remain behind in summer, or that we do not yet understand variation in this species.

At present I prefer to follow the nomenclature used by Godfrey (1986), rather than Dickerman (1987).

Common Redpoll, *Carduelis flammea*

Given that *C. f. holboelli* is of doubtful validity (AOU, 1957) and that there is considerable plumage variation in redpolls (Troy, 1985), it is doubtful that the single specimen supposedly of this form mentioned by Manning (1952) will withstand scrutiny.

Hoary Redpoll, *Carduelis hornemanni*

I tend to agree with Troy (1985) that Hoary and Common redpolls should be considered one variable species. There is gradation in plumage characteristics from one form to the other among specimens in ROM collections.

Birds that breed in Ontario are typically dark and small (like *flammea*) but, in the summer of 1985, lighter-coloured birds that seemed "hoary" were observed as breeders (Cadman, Eagles, and Helleiner, 1987). However, as none were collected, the racial identity will remain in doubt. During the winter very light and larger birds (*hornemanni*-like) occur in small numbers. There are also many others, occurring outside the breeding season, that cover the range of variation in subspecific forms.

American Goldfinch, *Carduelis tristis*

West of Lake Superior, birds are variable, and those typical of *tristis* or *pallida* may be found at the same location, indicating a zone of intergradation. In the extreme west, most are typical of *pallida*.

LITERATURE CITED

ABRAHAM, K. F.
    1984   Ross' Gull: new to Ontario. Ontario Birds 2: 116–119.

ALDRICH, J. W., A. J. DUVALL, and A. D. GEIS
    1958   Racial determination of origins of Mourning Doves in hunters' bags. Journal of Wildlife Management 22:71–75.

AMERICAN ORNITHOLOGISTS' UNION
    1957   Check-list of North American birds. 5th ed. Baltimore, American Ornithologists' Union. 691 pp.
    1983   Check-list of North American birds. 6th ed., Washington, D.C., American Ornithologists' Union. 877 pp.
    1985   Thirty-fifth supplement to the American Ornithologists' Union check-list of North American birds. Auk 102:680–686.
    1989   Thirty-seventh supplement to the American Ornithologists' Union check-list of North American birds. Auk 106:532–538.

AMES, J. H.
    1897   Rare birds taken in Toronto and vicinity. Auk 14:411–412.

ANKNEY, C. D., D. G. DENNIS, L. N. WISHARD, and J. E. SEEB
    1986   Low genetic variation between Black Ducks and Mallards. Auk 103:701–709.

ANONYMOUS
    1987   Jackdaws: chapter three (?). Birding 19:45.

ATKINSON, G. E.
    1894   *Polyborus cheriway* on Lake Superior. Biological Review of Ontario 1:9–10.

AVISE, J. C., C. D. ANKNEY, and W. S. NELSON
    1990   Mitochondrial gene trees and the evolutionary relationship of Mallard and Black ducks. Evolution 44:1109–1119.

AXTELL, H. H., P. BENHAM, and J. E. BLACK
    1977   Spotted Redshank sighted in Ontario. Canadian Field-Naturalist 91:90–91.

BAILLIE, J. L.
    1950a   The fall migration. Audubon Field Notes 4:12–14.
    1950b   The winter season. Audubon Field Notes 4:199–200.
    1951   The spring migration. Audubon Field Notes 5:253–254.
    1952   The fall migration. Audubon Field Notes 6:13–16.
    1955   The nesting season. Audubon Field Notes 9:375–377.
    1957   Recent additions to Ontario's bird list. Ontario Field Biologist 11:1–3.
    1958   Western Tanager an Ontario bird. Ontario Field Biologist 12:28–29.
    1961   More new Ontario breeding birds. Ontario Field Biologist 15:1–9.
    1964   Ontario's newest birds. Ontario Field Biologist 18:1–13.
    1965   Ontario's bird list; subtract one; add one. Ontario Field Biologist 19:41.

BAILLIE, J. L. and P. HARRINGTON
    1936   The distribution of breeding birds in Ontario. Transactions of the Royal Canadian Institute 21(1):1–50.

BAIN, M.
- 1978 First sighting of a Black Skimmer in Ontario. Ontario Field Biologist 32(2):33.
- 1986 Eurasian Jackdaw: new to Ontario. Ontario Birds 4:64–65.

BANKS, R.
- 1986 Subspecies of the Glaucous Gull, *Larus hyperboreus* (Aves: Charadriiformes). Proceedings of the Biological Society of Washington 99:149–159.

BARLOW, J. C.
- 1966 Status of the Wood Ibis, the Fulvous Tree Duck, and the Wheatear in Ontario. Canadian Field-Naturalist 80:183–186.
- 1967 Rufous Hummingbird in Ontario. Canadian Field-Naturalist 81:148–149.

BAXTER, T. S. H.
- 1985 The birding handbook: eastern Lake Superior. Wawa, Superior Lore. 133 pp.

BEARDSLEE, C. S. and H. D. MITCHELL
- 1965 Birds of the Niagara Frontier region. Bulletin of the Buffalo Society of Natural Sciences 22:1–478.

BELL, C. C.
- 1941 Unusual bird records for Kent County, Ontario. Canadian Field-Naturalist 55:13.

BELLROSE, F. C.
- 1976 Ducks, geese and swans of North America. 2nd ed. Harrisburg, Stackpole. 540 pp.

BENNETT, G.
- 1989 Winter birding across Canada. Birdfinding in Canada 9(3):5–15.

BERGTOLD, W. H.
- 1889 List of the birds of Buffalo and vicinity. Bulletin of the Buffalo Naturalist Field Club 1:3.

BLOKPOEL, H.
- 1987 California Gull. *In* Cadman, M. D., P. F. J. Eagles, and F. M. Helleiner, comps., Atlas of the breeding birds of Ontario. Waterloo, University of Waterloo Press, p. 564.

BRADSTREET, M. S. W. and J. D. McCRACKEN
- 1978 Avifaunal survey of St. Lawrence Islands National Park. Cornwall, Ontario, Parks Canada. 343 pp.

BRAUN, M. J. and M. B. ROBBINS
- 1986 Extensive protein similarity of the hybridizing chickadees *Parus atricapillus* and *P. carolinensis*. Auk 103:667–675.

BREWER, A. D.
- 1977 The birds of Wellington County. Guelph, Ontario, Guelph Field Naturalists Club Special Publication. 38 pp.

BREWER, D., M. A. W. LANE, and M. L. WERNAART
- 1984 Siberian Rubythroat: a new species to Canada. Ontario Birds 2:66–69.

BRODKORB, P.
- 1935 The name of the western race of Red-headed Woodpecker. Occasional Papers of the Museum of Zoology, University of Michigan 303:1–3.

BROWN, H. H.
- 1894 *Aestrelata hasitata* taken at Toronto. Biological Review of Ontario 1:11.

CADMAN, M. D., P. F. J. EAGLES, and F. M. HELLEINER, comps.
- 1987 Atlas of the breeding birds of Ontario. Waterloo, University of Waterloo Press. 617 pp.

CARPENTIER, A.G.
- 1987 Western Kingbird nesting in Rainy River District. Ontario Birds 7:33–34.
- 1990 Broad-billed Hummingbird: new to Ontario and Canada. Ontario Birds 8:34–37.

CHAMBERLAIN, M.
- 1887 A catalogue of Canadian birds. Saint John, H.A. McMillan. 119 pp.

CLARKE, C. H. D.
- 1943 The Wild Turkey in Ontario. Sylva 4:5–12, 24.
- 1954 The Bob-white in Ontario. Ontario Dept. of Lands and Forests, Technical Bulletin, Fish and Wildlife Series 2:1–10.

COADY, G.
- 1988 Ontario Bird Records Committee report for 1987. Ontario Birds 6:42–50.

COADY, G. and A. WORMINGTON
- 1989 Ontario Bird Records Committee report for 1988. Ontario Birds 7:43–54.

COLLIER, B. and J. CURSON
- 1988 Snowy Plover: new to Ontario. Ontario Birds 6:4–10.

COTTLE, T. J.
- 1859 Capture of two birds of unusual occurrence in Upper Canada. Canadian Journal, n.s. 4(23):388–389.

CURRY, R. and G. D. BRYANT
- 1987 Snowy Egret: a new breeding species for Ontario. Ontario Birds 5:64–67.

DENIS, K.
- 1958 Additions to breeding records—1957. Thunder Bay Field Naturalists Club Newsletter 12(2):15.
- 1976 Scott's Oriole near Thunder Bay, Ontario. Canadian Field-Naturalist 90:500–501.

DICK, J. A. and R. D. JAMES
- 1969 The Ground Dove in Canada. Canadian Field-Naturalist 83:405–406.

DICKERMAN, R. W.
- 1987 The "old northeastern" subspecies of Red Crossbill. American Birds 41:189–194.

DiLABIO, B. M. and J. M. BOUVIER
- 1986 Atlantic Puffin: new to Ontario. Ontario Birds 4:19–21.

DOW, D. D.
- 1962 First Canadian record of Virginia's Warbler. Auk 79:715.

DUNCAN, B. W.
- 1986 The occurrence and identification of Swainson's Hawk in Ontario. Ontario Birds 4:43–61.

ELDER, D. H.
- 1988 Great-tailed Grackle: new to Ontario. Ontario Birds 6:28–31.

FAZIO, V., D. SHEPHERD, and T. WOODROW
- 1985 A seasonal checklist of the birds of the Long Point area. Port Rowan, Long Point Bird Observatory.

FLEMING, J. H.
- 1900   Ontario bird notes. Auk 17:176.
- 1901   A list of birds of the districts of Parry Sound and Muskoka, Ontario. Auk 18:32–45.
- 1903   Recent records of the Wild Pigeon. Auk 20:66.
- 1906   Birds of Toronto, Ontario. Part 1, water birds. Auk 23:437–453.
- 1907   Birds of Toronto, Ontario. Part 2, land birds. Auk 24:71–89.
- 1912   The Ancient Murrelet (*Synthliboramphus antiquus*) in Ontario. Auk 29:387–388.
- 1913   Ontario bird notes. Auk 30:225–228.

FRANCIS, G.
- 1984   Habitat relationships, management and interpretation of the bird communities of the eastern sector of Georgian Bay Islands National Park. Cornwall, Parks Canada. 2 vols. 515 pp.
- 1985   The birds of the Tobermory Islands unit. Cornwall, Parks Canada. 300 pp.

FRIEDMANN, H.
- 1950   The birds of North and Middle America. Part 11. United States National Museum, Bulletin 50:1–793.

GASTON, A. J. and R. DECKER
- 1985   Interbreeding of Thayer's Gull, *Larus thayeri*, and Kumlien's Gull, *Larus glaucoides kumlieni*, on Southampton Island, Northwest Territories. Canadian Field-Naturalist 99:257–259.

GAWN, M.
- 1987   Sulphur-bellied Flycatcher: new to Ontario and Canada. Ontario Birds 5:87–93.

GODFREY, W. E.
- 1951   The Nevada Cowbird at James Bay, Ontario. Canadian Field-Naturalist 65:46.
- 1973   A possible shortcut spring migration route of the Arctic Tern to James Bay, Canada. Canadian Field-Naturalist 87:51–52.
- 1976   Audubon's Shearwater, a new species for Canada. Canadian Field-Naturalist 90:494.
- 1986   The birds of Canada. Ottawa, National Museums of Canada. 595 pp.

GOLLOP, J. B., T. W. BARRY, and E. H. IVERSEN
- 1986   Eskimo Curlew. A vanishing species? Saskatchewan Natural History Society Special Publication 17:1–160.

GOODWIN, C. E.
- 1962   Worth noting. Bulletin of the Federation of Ontario Naturalists 98:18–27.
- 1966   The spring migration. Audubon Field Notes 20:501–504.
- 1968   The winter season. Audubon Field Notes 22:434–436.
- 1969   The spring migration. Audubon Field Notes 23:583–586.
- 1970   The fall migration. Audubon Field Notes 24:38–43.
- 1971a The fall migration. American Birds 25:49–54.
- 1971b The winter season. American Birds 25:570–575.
- 1972a Ontario Ornithological Records Committee report for 1972. Ontario Field Biologist 26:35–37.
- 1972b The winter season. American Birds 26:597–601.

1973a   The fall migration. American Birds 27:49-56.
1973b   The winter season. American Birds 27:608-611.
1974a   The winter season. American Birds 28:632-635.
1974b   The spring migration. American Birds 28:794-798.
1975   The spring migration. American Birds 29:843-848.
1976a   The winter season. American Birds 30:711-715.
1976b   The spring migration. American Birds 30:832-836.
1976c   Ontario Ornithological Records Committee report for 1975. Ontario Field Biologist 30:35-37.
1977a   Ontario Ornithological Records Committee report for 1977. Ontario Field Biologist 31:13-16.
1977b   The nesting season. American Birds 31:1131-1135.
1978   The spring migration. American Birds 32:997-1001.
1979a   The winter season. American Birds 33:276-279.
1979b   Ontario Ornithological Records Committee report for 1978. Ontario Field Biologist 33:24-25.
1980a   The winter season. American Birds 34:267-270.
1980b   The spring migration. American Birds 34:770-773.
1981   The winter season. American Birds 35:295-298.

GUNN, W. W. H.
1957   The spring migration. Audubon Field Notes 11:342-344.

HAIG, S. M. and L. W. ORING
1988   Distribution and dispersal in the Piping Plover. Auk 105:630-638.

HARRINGTON, P.
1915   Ontario, 1914 nests. Oologist 32:99.
1939   Kirtland's Warbler in Ontario. Jack-Pine Warbler 17:95-97.

HARRIS, C. G.
1983   Sight record of a Golden-crowned Sparrow (*Zonotrichia atricapilla*) in Ontario. Ontario Birds 1:70-71.

HOPE, C. E.
1949   First occurrence of the Black Vulture in Ontario. Auk 66:81-82.

HOWE, L.
1973   Rock Wren found at Ear Falls. Thunder Bay Field Naturalists Club Newsletter 27:50-51.

HUGHES, A. E.
1971   A Gray Kingbird in eastern Ontario. Blue Bill 18:45.

HUSSELL, D. J. T. and M. J. PORTER
1977   Fieldfare in Ontario. Canadian Field-Naturalist 91:91-92.

JAMES, R. D.
1978   Nesting of the House Finch (*Carpodacus mexicanus*) in Ontario. Ontario Field Biologist 32:30-32.
1982   Ontario Ornithological Records Committee report for 1981. Ontario Field Biologist 36:16-18.
1983   Ontario Bird Records Committee report for 1982. Ontario Birds 1:7-15.
1984a   The breeding bird list for Ontario: additions and comments. Ontario Birds 2:24-29.
1984b   Ontario Bird Records Committee report for 1983. Ontario Birds 2:53-65.

JAMES, R. D., P. L. McLAREN, and J. C. BARLOW
  1976 Annotated checklist of the birds of Ontario. Life Sciences Miscellaneous Publications, Toronto, Royal Ontario Museum. 75 pp.

JARVIS, J.
  1965 A possible occurrence of the Carolina Chickadee (*Parus carolinensis*) in southwestern Ontario. Ontario Field Biologist 19:42.

KELLEY, A. H.
  1978 Birds of southeastern Michigan and southwestern Ontario. Cranbrook Institute of Science, Bulletin 57:1–99.

LAMEY, J.
  1981 Unusual records of birds for Ontario's Rainy River District. Ontario Bird Banding 14:38–42.

LANYON, E. and J. BULL
  1967 Identification of Connecticut, Mourning and MacGillivray's Warblers. Bird Banding 38:187–194.

LLOYD, H.
  1923 The birds of Ottawa. Canadian Field-Naturalist 37:101–105.

LONG, R. C.
  1968 First occurrence of Cassin's Sparrow in Canada. Ontario Field Biologist 22:34.

LUMSDEN, H. G.
  1959 Mandt's Black Guillemot breeding on Hudson Bay coast of Ontario. Canadian Field-Naturalist 73:54–55.
  1966 The Prairie Chicken in southwestern Ontario. Canadian Field-Naturalist 80:33–45.
  1971 The status of the Sandhill Crane in northern Ontario. Canadian Field-Naturalist 85:285–293.
  1984a A swan song with a difference. Seasons 24(2):18–23, 39.
  1984b The pre-settlement breeding distribution of Trumpeter, *Cygnus buccinator* and Tundra Swans, *C. columbianus* in eastern Canada. Canadian Field-Naturalist 98:415–424.
  1987a Brant. *In* Cadman, M. D., P. F. J. Eagles, and F. M. Helleiner, comps., Atlas of the breeding birds of Ontario. Waterloo, University of Waterloo Press, p. 563.
  1987b Greater Prairie-Chicken. *In* Cadman, M.D., P. F. J. Eagles, and F. M. Helleiner, comps., Atlas of the breeding birds of Ontario. Waterloo, University of Waterloo Press, p. 563.

LUNN, J.
  1961 An American Oystercatcher on the Great Lakes. Ontario Field Biologist 15:32–33.

MacFAYDEN, C. J.
  1945 Breeding of *Tyrannus verticalis* in Ontario. Canadian Field-Naturalist 59:67.

MACK, A. L., F. B. GILL, R. COLUBUN, and C. SPOLSKY
  1986 Mitochondrial DNA: a source of genetic markers for studies of similar passerine bird species. Auk 103:676–681.

MacKENZIE, H. N.
    1968   A possible Fieldfare observation near Ottawa, Ontario. Canadian Field-Naturalist 82:51.

MACOUN, J. and J. M. MACOUN
    1909   Catalogue of Canadian birds. Ottawa, Canada Dept. of Mines, Geological Survey Branch. 761 pp.

MACOUN, W. T.
    1898   Bird notes for June. Ottawa Naturalist 12:89.

MANNING, T. H.
    1952   Birds of the west James Bay and southern Hudson Bay coasts. National Museum of Canada, Bulletin 125:1–114.

McCRACKEN, J. D.
    1987   The breeding birds of Haldimand-Norfolk. *In* Gartshore, M. E., D. A. Sutherland, and J. D. McCracken, Natural areas inventory of the Regional Municipality of Haldimand-Norfolk. Simcoe, Norfolk Field Naturalists. Vol. 2, Part 5.
    1988   An enigmatic case for the breeding of the Kentucky Warbler in Canada. Ontario Birds 6:101–105.

McCRACKEN, J. D., M. S. W. BRADSTREET, and G. L. HOLROYD
    1981   Breeding birds of Long Point, Lake Erie: a study in community succession. Canadian Wildlife Service, Report Series 44:1–72.

McILWRAITH, T.
    1886   The birds of Ontario. Hamilton, Hamilton Association. 303 pp.
    1894   The birds of Ontario. Toronto, William Briggs. 426 pp.

McRAE, R. D.
    1982   Birds of Presqu'ile, Ontario. Toronto, Ontario Ministry of Natural Resources. 74 pp.
    1985   Mongolian Plover: new to Canada. Ontario Birds 3:18–23.

McRAE, R. D. and W. H. HUTCHISON
    1983   A record of the Yellow-throated Warbler from Moosonee. Ontario Birds 1:16–17.

MILLS, A.
    1981   A cottager's guide to the birds of Muskoka and Parry Sound. Guelph, Published by the author. 209 pp.

MITCHELL, H. D. and R. F. ANDRLE
    1970   Birds of the Niagara Frontier Region supplement. Bulletin of the Buffalo Society of Natural Sciences 22(suppl.):1–10.

MITCHELL, M. H.
    1935   The Passenger Pigeon in Ontario. Contributions of the Royal Ontario Museum of Zoology 7:1–181.

MORRISON, R. I. G.
    1980   First specimen record of the Little Stint (*Caladris minuta*) for North America. Auk 97:627–628.

MOUNTJOY, J. D. and R. D. McRAE
    1983   An Ash-throated Flycatcher (*Myiarchus cinerascens*) at Whitby. Ontario Birds 1:64–66.

NASH, C. W.
    1894   Black Rail in Ontario. Biological Review of Ontario 1:13.

NICHOLSON, J. C.
    1981   The birds of Manitoulin Island and adjacent islands within Manitoulin District. Sudbury, Published by the author. 205 pp.

NOL, E.
    1983   The first substantiated record of the Painted Bunting (*Passerina ciris*) in Ontario. Ontario Birds 1:33–34.

NORTH, G.
    1956   Noteworthy bird records. Wood Duck 9(8):7.

OUELLET, H. R.
    1977   Biosystematics and ecology of *Picoides villosus* (L.) and *P. pubescens* (L.), (Aves: Picidae). Ph.D. Thesis, McGill University.

PALMER, R. S.
    1976   Handbook of North American Birds. Vols. 2 and 3. New Haven, Yale University Press. 533 & 568 pp.

PARKES, K. C.
    1988   The Ontario specimen of Carolina Chickadee. Ontario Birds 6:111–114.

PAYNE, R. B.
    1983   A distributional checklist of the birds of Michigan. Miscellaneous Publications, Museum of Zoology, University of Michigan. 164:1–71.

PECK, G. K.
    1966   First published breeding record of Mute Swan for Ontario. Ontario Field Biologist 20:43.
    1972   Birds of the Cape Henrietta Maria region, Ontario. Canadian Field-Naturalist 86:333–349.

PECK, G. K. and R. D. JAMES
    1983   The breeding birds of Ontario: nidiology and distribution. Volume 1: Nonpasserines. Life Sciences Miscellaneous Publications. Toronto, Royal Ontario Museum. 321 pp.
    1987   The breeding birds of Ontario: nidiology and distribution. Volume 2: Passerines. Life Sciences Miscellaneous Publications. Toronto, Royal Ontario Museum. 387 pp.

PERUNIAK, S.
    1971   The birds of the Atikokan area, Rainy River District, Ontario. Part 2. Ontario Field Biologist 25:15–33.

PITELKA, F. A.
    1950   Geographic variation and the species problem in the Shorebird genus *Limnodromus*. University of California Publications in Zoology 50:1–108.

PREVEC, R.
    1985   Archaeological evidence of the Carolina Parakeet in Ontario. Ontario Birds 3:24–28.

QUILLIAM, H. R.
    1973   History of the birds of Kingston, Ontario. 2nd ed. Kingston, Kingston Field Naturalists. 209 pp.

RAND, A. L.
　1948　Variation in the Spruce Grouse in Canada. Auk 65:33–40.
RAYNER, W. J.
　1988　First nest record of White-eyed Vireo in Ontario. Ontario Birds 6:114–116.
REINECKE, O.
　1916a　Rarities. Oologist 33:13.
　1916b　Passenger Pigeon. Oologist 33:126.
RIDGWAY, R.
　1914　The birds of North and Middle America. Bulletin of the United States National Museum 50(6):1–882.
ROBBINS, C. S.
　1960　Status of the Bald Eagle, summer of 1959. United States Fish and Wildlife Service, Wildlife Leaflet 418:1–8.
ROBBINS, M. B., M. J. BRAUN, and E. A. TOBEY
　1986　Morphological and vocal variation across a contact zone between the chickadees *Parus atricapillus* and *P. carolinensis*. Auk 103:655–666.

SADLER, D. C.
　1983　Our heritage of birds: Peterborough County in the Kawarthas. Peterborough, Peterborough Field Naturalists. 190 pp.
SALOMONSEN, F.
　1944　The Atlantic Alcidae; the seasonal and geographic variation of the auks inhabiting the Atlantic Ocean and adjacent waters. Goteborgs Kungl. Vetenskaps—Och Vitterhets—Samhalles Handlingar, Sjatte Foljden, Ser. B, Bd. 3(5):1–138.
SANDILANDS, A. P.
　1984　Annotated checklist of the vascular plants and vertebrates of Luther Marsh, Ontario. Special Publication 2, Ontario Field Biologist. 134 pp.
SAUNDERS, W. E.
　1919　Bachman's Sparrow an addition to the Canadian fauna. Canadian Field-Naturalist 33:118.
SAUNDERS, W. E. and E. M. S. DALE
　1933　History and list of birds of Middlesex County, Ontario. Transactions of the Royal Canadian Institute 19:161–248.
SCHUELER, F. W., D. H. BALDWIN, and J. D. RISING
　1974　The status of birds at selected sites in northern Ontario. Canadian Field-Naturalist 88:141–150.
SHEPPARD, R. W.
　1944　Sycamore Warbler in Ontario. Auk 61:469.
　1960　Bird life of Canada's Niagara frontier. Privately published. 50 pp. Mimeographed.
　1970　Bird life of Canada's Niagara frontier. Niagara Falls Nature Club, Special Publication 3:1–85.
SHORTT, T. M. and C. E. HOPE
　1943　White-winged Dove in Ontario. Auk 60:449–450.
SKEEL, M. and S. BONDRUP-NIELSEN
　1978　Avifauna survey of Pukaskwa National Park. Cornwall, Parks Canada. 259 pp.

SNELL, R. R.
- 1989 Status of *Larus* gulls at Home Bay, Baffin Island. Colonial Waterbirds 12:12–23.

SNYDER, L. L.
- 1926 First record of the Lark Bunting for Ontario. Auk 43:375.
- 1938 A faunal investigation of western Rainy River District, Ontario. Transactions of the Royal Canadian Institute 22:157–213.
- 1941 The birds of Prince Edward County, Ontario. University of Toronto Studies, Biological Series 48:25–92.
- 1953 Summer birds of western Ontario. Transactions of the Royal Canadian Institute 30:47–95.
- 1954 Cassin's Kingbird in Canada. Auk 71:209.
- 1961 On an unnamed population of the Great Horned Owl. Life Sciences Division, Royal Ontario Museum Contribution 54:1–7.

SPEIRS, D. H.
- 1984 First breeding record of Kirtland's Warbler in Ontario. Ontario Birds 2:80–84.

SPEIRS, J. M.
- 1958 Worth noting. Bulletin of the Federation of Ontario Naturalists 82:25–33.
- 1959 Worth noting. Bulletin of the Federation of Ontario Naturalists 85:22–31.
- 1960a The fall migration. Audubon Field Notes 14:30–35.
- 1960b The winter season. Audubon Field Notes 14:305–309.
- 1960c Worth noting. Bulletin of the Federation of Ontario Naturalists 90:13–24.
- 1985 Birds of Ontario. Volume 2. Toronto, Natural Heritage/Natural History. 986 pp.

SPEIRS, J. M. and E. PEGG
- 1972 First record of Painted Redstart (*Setophaga picta*) for Canada. Auk 89:898.

SPRAGUE, R. T. and R. D. WEIR
- 1984 The birds of Prince Edward County. Kingston, Kingston Field Naturalists. 190 pp.

STIRRETT, G.
- 1941 The Scarlet Ibis and other waders at Point Pelee National Park. Canadian Field-Naturalist 55:13.
- 1973a The spring birds of Point Pelee National Park. Ottawa, Information Canada. 32 pp.
- 1973b The autumn birds of Point Pelee National Park. Ottawa, Information Canada. 36 pp.

STORER, R. W.
- 1952 A comparison of variation, behavior and evolution in the sea bird genera *Uria* and *Cepphus*. University of California Publications in Zoology 52:121–222.

SWAINSON, W. and J. RICHARDSON
- 1831 Fauna Boreali-Americana. Part II. The Birds. London. 524 pp.

TACHA, T. C., P. A. VOHS, and W. D. WARDE
- 1985 Morphometric variation of Sandhill Cranes from mid-continental North America. Journal of Wildlife Management 49:246–250.

TAVERNER, P. A.
  1934   The Madeira Petrel in Ontario. Auk 51:77.
TEBBEL, P. D.
  1981   The status, distribution and nesting ecology of Sandhill Cranes in the Algoma District of Ontario. MSc. Thesis, University of Western Ontario. 66 pp.
THOMPSON, E. E.
  1890   Proceedings of the ornithological subsection of the biological section of the Canadian Institute. Proceedings of the Canadian Institute, 3rd series, 7(2):181–202.
TODD, W. E. C.
  1940   Eastern races of the Ruffed Grouse. Auk 57:390–397.
  1947   A new name for *Bonasa umbellus canescens* Todd. Auk 64:326.
TONER, G. C.
  1940   Leach's Petrel in Ontario. Wilson Bulletin 52:124.
TROY, D. M.
  1985   A phenetic analysis of the redpolls *Carduelis flammea flammea* and *C. hornemanni exilipes.* Auk 102:82–96.
TUCK, L. M.
  1968   Dowitcher breeding in Ontario. Ontario Field Biologist 21:39.
VAN TYNE, J.
  1950   Old record of *Anhinga anhinga* taken on St. Mary's River, Ontario. Auk 67:508–509.
WALKINSHAW, L. H.
  1965   A new Sandhill Crane from central Canada. Canadian Field-Naturalist 79:181–184.
WEIR, R. D.
  1970   Possible House Finch. Blue Bill 17:60–61.
  1974   Spring and summer seasons. Blue Bill 21:53–58.
  1982   The winter season. American Birds 36:289–291.
  1983a The autumn migration. American Birds 37:173–177.
  1983b The nesting season. American Birds 37:982–985.
  1985a The winter season. American Birds 39:161–164.
  1985b The nesting season. American Birds 39:905–909.
  1986a The autumn migration. American Birds 40:104–109.
  1986b The winter season. American Birds 40:274–277.
  1986c The spring migration. American Birds 40:462–467.
  1987a The winter season. American Birds 41:275–279.
  1987b The nesting season. American Birds 41:1429–1432.
  1988a The winter season. American Birds 42:256–261.
  1988b The spring migration. American Birds 42:426–430.
  1988c The nesting season. American Birds 42:1281–1286.
  1989a The autumn migration. American Birds 43:94–99.
  1989b The birds of the Kingston region. Kingston, Kingston Field Naturalists. 608 pp.
  1989c The winter season. American Birds 43:307–311.
  1989d The spring migration. American Birds 43:470–475.

    1990a   The spring migration. American Birds 44:418–423.

    1990b   The nesting season. American Birds 44:(in press).

WEIR, R. D. and H. R. QUILLIAM

    1980   Supplement to History of the Birds of Kingston, Ontario. Special Publication, Kingston Field Naturalist. 40 pp.

WEIR, R. D., H. QUILLIAM, and R. NORMAN

    1972   First record of Chestnut-collared Longspur in Ontario. Canadian Field-Naturalist 86:382–383.

WHITE, C. M.

    1968   Diagnosis and relationships of the North American tundra-inhabiting Peregrine Falcons. Auk 85:179–191.

WILCOX, L.

    1959   A twenty year banding study of the Piping Plover. Auk 76:129–152.

WOODFORD, J.

    1963   The spring migration. Audubon Field Notes 17:399–401.

    1964   The fall migration. Audubon Field Notes 18:28–30.

WORMINGTON, A.

    1982   The rare breeding birds of Point Pelee National Park—a status report, 1982 census and recommendations for resource management. Cornwall, Parks Canada. 30 pp.

    1984a   Seventh annual (1984) spring migration report. Point Pelee National Park and vicinity. Cornwall, Parks Canada. 20 pp.

    1984b   An observation of "Richardson's" Merlin in Ontario. Ontario Birds 4:62–64.

    1985   Ontario Bird Records Committee report for 1984. Ontario Birds 3:2–17.

    1986   Ontario Bird Records Committee report for 1985. Ontario Birds 4:3–18.

    1987   Ontario Bird Records Committee report for 1986. Ontario Birds 5:42–63.

WORMINGTON, A. and R. H. CURRY

    1990   Ontario Bird Records Committee report for 1989. Ontario Birds 8:4–33.

WYETT, W. R.

    1966   First Ontario specimen of Mountain Bluebird collected at Point Pelee. Ontario Field Biologist 20:42.

ADDITIONAL REFERENCES

Titles listed here, although not quoted, were consulted for information, some many times. Most contributed directly to the compilation of this checklist, and others form an historical basis to more recent works.

ABRAHAM, K. F. and G. H. FINNEY
    1986   Eiders of the eastern Canadian Arctic. *In* Reed, A., ed., Eider ducks in Canada. Canadian Wildlife Service Report Series 47:55–73.

ALDRICH, J. W.
    1944   Geographic variation of Bewick's Wrens in the eastern United States. Occasional Papers of the Museum of Zoology, Louisiana State University 18:305–309.

ALLIN, A. E.
    1940   Vertebrate fauna of Darlington Township, Durham County. Transactions of the Royal Canadian Institute 49:83–118.

BAILLIE, J. L.
    1947   The summer birds of Sudbury District, Ontario. Contributions of the Royal Ontario Museum of Zoology 28:1–32.

BAILLIE, J. L. and P. HARRINGTON
    1937   The distribution of breeding birds in Ontario. Transactions of the Royal Canadian Institute 21(2):199–283.

BAILLIE, J. L. and C. E. HOPE
    1943   The summer birds of the northeast shore of Lake Superior, Ontario. Contributions of the Royal Ontario Museum of Zoology 23:1–27.

BREMNER, R. M.
    1949   Observations of the birds of the Casummit-Birch Lakes region of northwest Ontario. Canadian Field-Naturalist 63:161–165.

BROOMAN, R. C.
    1954   The birds of Elgin County. St. Thomas, Gilbert Press. 41 pp.

BROWN, N. R.
    1947   Observations on the birds of the Petawawa Military Reserve and surrounding district, Renfrew County, Ontario. Canadian Field-Naturalist 61:47–55.

CALVERT, E. W.
    1925   A preliminary list of the birds of the Lindsay District, Ontario. Canadian Field-Naturalist 39:49–51, 72–74.

DEAR, L. S.
    1940   Breeding birds of the region of Thunder Bay, Lake Superior, Ontario. Transactions of the Royal Canadian Institute 23:119–143.

DENIS, K.
    1961   Birds of the Canadian Lakehead area. Thunder Bay Field Naturalists Club, Supplement 2. 8 pp. Mimeographed.

DEVITT, O. E.
1967 The birds of Simcoe County, Ontario. Barrie, Brereton Field Naturalists Club. 192 pp.

GOODWILL, J. E. V.
1942 The summer birds of the Madsen Area, Patricia District, Ontario. Canadian Field-Naturalist 56:131–133.

KELLEY, A. H.
1983 Birds of S.E. Michigan and S.W. Ontario: notes on the years 1975–1981. Jack-Pine Warbler 61:3–12.

KELLEY, A. H., D. S. MIDDLETON, and W. P. NICKELL
1963 Birds of the Detroit-Windsor area. Cranbrook Institute of Science, Bulletin 45:1–119.

LEE, D.
1978 An annotated list of the birds of the Big Trout Lake area, Kenora District. Ontario Field Biologist 32:17–36.

LEWIS, H. F. and H. S. PETERS
1941 Notes on birds of the James Bay region in the autumn of 1940. Canadian Field-Naturalist 55:111–117.

LLOYD, H. and R. G. LANNING
1948 Observations on the birds of Renfrew County, Ontario. Canadian Field-Naturalist 62:47–65.

LONG, R. C.
1965– An annotated list of the birds of Pickering Beach. 7 parts. Ontario Field
1972 Biologist 19:26–35; 20:25–34; 21:14–29; 22:8–24; 23:14–23; 24:19,21,22; 26:38–45.

MacLULICH, D. A.
1938 Birds of Algonquin Provincial Park, Ontario. Contributions of the Royal Ontario Museum of Zoology 13:1–47.

McKEOUGH, G. T. and J. S. SMITH
1924 Some remarks on birds, with a list of the birds of the County of Kent. Kent Historical Society Papers and Addresses 6:49–74.

McLAREN, M. A. and P. L. McLAREN
1981 Relative abundances of birds in boreal and subarctic habitats of northwestern Ontario and northeastern Manitoba. Canadian Field-Naturalist 95:418–427.

McLAREN, P. L. and M. A. McLAREN
1981 Bird observations in northwestern Ontario, 1976–77. Ontario Field Biologist 35:1–6.

MENDALL, H. L.
1986 Identification of eastern races of the Common Eider. *In* Reed, A., ed., Eider ducks in Canada. Canadian Wildlife Service Report Series 47:82–88.

MITCHELL, M. H.
1929 Summer birds of Miner's Bay and vicinity, Haliburton County, Ontario. Canadian Field-Naturalist 43:147–152.

MORDEN, J. A. and W. E. SAUNDERS
  1882   List of the birds of western Ontario. Canadian Sportsman and Naturalist, 2:183–187, 192–194.

NASH, C. W.
  1908   Check list of the birds of Ontario. *In* Nash, C.W., Vertebrates of Ontario. Toronto, Dept. of Education, pp. 7–82.

NICHOLSON, J. C.
  1972   The birds of Manitoulin Island. Sudbury, Published by the author. 46 pp.

PALMER, R. S.
  1962   Handbook of North American birds. Vol. 1. New Haven, Yale University Press. 567 pp.

PERUNIAK, S.
  1969   The birds of the Atikokan area, Rainy River District, Ontario. Ontario Field Biologist 23:35–38; 25:15–33.

PETERSON, J. M. C.
  1985   Birds of Little Sachigo Lake and Thorne-Sachigo Rivers, Ontario. Ontario Birds 3:87–99.

PRATER, A. J., J. H. MARCHANT, and J. VUORINEN
  1977   Guide to the identification and ageing of Holarctic waders. British Trust for Ornithology Guide 17. 168 pp.

REED, A. and A. J. ERSKINE
  1986   Populations of the Common Eider in eastern North America: their size and status. *In* Reed, A., ed., Eider ducks in Canada. Canadian Wildlife Service Report Series 47:156–162, 170–175.

RICKER, W. E. and C. H. D. CLARKE
  1939   The birds of the vicinity of Lake Nipissing, Ontario. Contributions of the Royal Ontario Museum of Zoology 16:1–25.

RYFF, A. J.
  1987   Smith's Longspur: a case of neglect. Ontario Birds 5:2–20.

SMITH, W. J.
  1957   Birds of the Clay Belt of Northern Ontario and Quebec. Canadian Field-Naturalist 71:163–181.

SNYDER, L. L.
  1928   The summer birds of Lake Nipigon. Transactions of the Royal Canadian Institute 16:251–277.
  1930   A faunal investigation of King Township, York County, Ontario. Transactions of the Royal Canadian Institute 17:183–202.
  1935   A study of the Sharp-tailed Grouse. University of Toronto Studies, Biological Series 40:1–66.
  1942   Summer birds of the Sault Ste. Marie region, Ontario. Transactions of the Royal Canadian Institute 24:121–153.
  1953   Summer birds of western Ontario. Transactions of the Royal Canadian Institute 30:47–95.

SNYDER, L. L. and E. D. LAPWORTH
- 1953 A comparative study of adults of two Canadian races of red-wings. Canadian Field-Naturalist 67:143–147.

SNYDER, L. L. and T. M. SHORTT
- 1946 Variation in *Bonasa umbellus*, with particular reference to the species in Canada east of the Rockies. Contributions of the Royal Ontario Museum of Zoology 27:118–133.

SNYDER, L. L. and E. M. WALKER
- 1928 A faunal investigation of the Lake Abitibi region, Ontario. University of Toronto Studies, Biological Series 32:1–46.

SPEIRS, D. H. and J. M. SPEIRS
- 1947 Birds of the vicinity of North Bay, Ontario. Canadian Field-Naturalist 61:23–38.

SPEIRS, J. M.
- 1973–1979 Birds of Ontario County. Toronto, Federation of Ontario Naturalists. 6 vols. 464 pp.

SWALES, B. H. and P. A. TAVERNER
- 1906 Remarks on the summer birds of Lake Muskoka, Ontario. Wilson Bulletin 13:60–68.

TAVERNER, P. A. and B. H. SWALES
- 1907–1908 The birds of Point Pelee. Wilson Bulletin 19:37–54, 82–99; 20:79–96, 107–129.

TODD, W. E. C.
- 1963 Birds of the Labrador Peninsula. Toronto, University of Toronto Press. 819 pp.

TONER, G. C., W. E. EDWARDS, and M. W. CURTIS
- 1942 Birds of Leeds County, Ontario. Canadian Field-Naturalist 56:8–12, 21–24, 34–44, 50–56.

TOZER, R. G. and J. M. RICHARDS
- 1974 Birds of the Oshawa-Lake Scugog Region, Ontario. Oshawa, Published by the authors. 384 pp.

WENDT, J. S. and E. SILIEFF
- 1986 The kill of eiders and other sea ducks by hunters in eastern Canada. *In* Reed, A., ed., Eider ducks in Canada. Canadian Wildlife Service Report Series 47:147–154.

WETMORE, A.
- 1936 A new race of the Song Sparrow from the Appalachian region. Smithsonian Miscellaneous Collections 95(17):1–3.

WILLIAMS, M. Y.
- 1920 Notes on the fauna of the Moose River and the Mattagami and Abitibi tributaries. Canadian Field-Naturalist 34:121–126.
- 1921 Notes on the fauna of lower Pagwachuan, lower Kenogami and lower Albany rivers of Ontario. Canadian Field-Naturalist 35:94–98.
- 1942 Notes on the fauna of Bruce Peninsula, Manitoulin and adjacent islands. Canadian Field-Naturalist 56:60–62, 70–81.

WORMINGTON, A. and R. D. JAMES
- 1984 Ontario Bird Records Committee, checklist of the birds of Ontario. Ontario Birds 2:13–23.

WRIGHT, A. H. and S. E. R. SIMPSON
- 1920 The vertebrates of the Otter Lake region, Dorset, Ontario. Canadian Field-Naturalist 34:141–145, 161–168.

INDEX TO COMMON AND SCIENTIFIC NAMES

Accipiter
 cooperii, 27
 gentilis, 27–28
 striatus, 27
Actitis
 macularia, 36
Aechmophorus
 occidentalis, 15
Aegolius
 acadicus, 49
 funereus, 49
Agelaius
 phoeniceus, 81
Aimophila
 aestivalis, 75
 cassinii, 75
Aix
 sponsa, 21
Alca
 torda, 45
Alle
 alle, 45
Ammodramus
 bairdii, 77
 caudacutus, 78, 95
 henslowii, 78
 leconteii, 78
 savannarum, 77, 95
Anas
 acuta, 22
 americana, 23
 bahamensis, 87
 clypeata, 22
 crecca, 21
 cyanoptera, 22
 discors, 22, 89
 penelope, 22
 platyrhynchos, 22
 rubripes, 21, 88–89
 strepera, 22
Anhinga, 17
Anhinga
 anhinga, 17

Ani,
 Groove-billed, 47
Anser
 albifrons, 20
 fabalis, 87
Anthus
 rubescens, 64
 spragueii, 64
Aphelocoma
 coerulescens, 87
Aquila
 chrysaetos, 29
Archilochus
 colubris, 51
Ardea
 herodias, 17
Arenaria
 interpres, 37, 90
Asio
 flammeus, 49
 otus, 49
Athene
 cunicularia, 48
Avocet,
 American, 35
Aythya
 affinis, 24
 americana, 23
 collaris, 23
 fuligula, 23
 marila, 23
 valisineria, 23

Bartramia
 longicauda, 36
Bittern,
 American, 17
 Least, 17
Blackbird,
 Brewer's, 82
 Eurasian, 63
 Red-winged, 81
 Rusty, 82
 Yellow-headed, 81

Bluebird,
  Eastern, 61
  Mountain, 61–62
Bobolink, 80–81
Bobwhite,
  Northern, 32
Bombycilla
  cedrorum, 64
  garrulus, 64
Bonasa
  umbellus, 31, 90
Botaurus
  lentiginosus, 17
Brambling, 83
Brant, 7, 20
Branta
  bernicla, 20
  canadensis, 21, 88
  leucopsis, 21
Bubo
  virginianus, 48, 91–92
Bubulcus
  ibis, 18
Bucephala
  albeola, 25
  clangula, 25
  islandica, 25
Bufflehead, 25
Bunting,
  Indigo, 74
  Lark, 77
  Lazuli, 74
  Painted, 75
  Snow, 80
Buteo
  jamaicensis, 28, 89
  lagopus, 28
  lineatus, 28
  platypterus, 28
  swainsoni, 28
Butorides
  striatus, 18

Calamospiza
  melanocorys, 77
Calcarius
  lapponicus, 80
  ornatus, 80
  pictus, 80

Calidris
  acuminata, 38
  alba, 37
  alpina, 39
  bairdii, 38
  canutus, 37
  ferruginea, 39
  fuscicollis, 38
  himantopus, 39
  maritima, 39
  mauri, 38
  melanotos, 38
  minuta, 38
  minutilla, 38
  pusilla, 37–38
  ruficollis, 38
Canvasback, 23
Caprimulgus
  carolinensis, 50
  vociferus, 50
Caracara,
  Crested, 29
Cardinal,
  Northern, 74
Cardinalis
  cardinalis, 74
  sinuatus, 87
Carduelis
  cannabina, 88
  carduelis, 88
  flammea, 84, 96
  hornemanni, 85, 96
  pinus, 85
  psaltria, 85
  spinus, 88
  tristis, 85, 96
Carpodacus
  mexicanus, 84
  purpureus, 83–84
Casmerodius
  albus, 17
Catbird,
  Gray, 63
Catharacta
  skua, 41
Cathartes
  aura, 26
Catharus
  fuscescens, 62, 93

guttatus, 62
minimus, 62, 93
ustulatus, 62
Catoptrophorus
   semipalmatus, 36
Cepphus
   grylle, 45–46, 91
Certhia
   americana, 59
Ceryle
   alcyon, 51
Chaetura
   pelagica, 50
Charadrius
   alexandrinus, 34
   melodus, 34, 90
   mongolus, 34
   semipalmatus, 34
   vociferus, 35
   wilsonia, 34
Chat,
   Yellow-breasted, 73
Chen
   caerulescens, 20, 88
   rossii, 20
Chickadee,
   Black-capped, 58
   Boreal, 58–59
   Carolina, 6, 58, 93
   Mountain, 58
Chlidonias
   niger, 45
Chondestes
   grammacus, 77
Chordeiles
   acutipennis, 50
   minor, 50
Chuck-will's-widow, 50
Circus
   cyaneus, 27
Cistothorus
   palustris, 60
   platensis, 60
Clangula
   hyemalis, 24
Coccothraustes
   vespertinus, 85
Coccyzus
   americanus, 47

erythropthalmus, 47
Colaptes
   auratus, 52, 92
Colinus
   virginianus, 32
Columba
   fasciata, 46, 91
   livia, 46
Columbina
   passerina, 47
Contopus
   borealis, 53
   sordidulus, 53
   virens, 53
Conuropsis
   carolinensis, 47
Coot,
   American, 33
Coragyps
   atratus, 26
Cormorant,
   Double-crested, 16
   Great, 16
Corvus
   brachyrhynchos, 58
   corax, 58
   cryptoleucus, 87
   monedula, 57
   ossifragus, 58
Coturnicops
   noveboracensis, 32
Cowbird,
   Brown-headed, 82, 95
Crane,
   Sandhill, 33, 90
   Whooping, 33
Creeper,
   Brown, 59
Crossbill,
   Red, 84, 96
   White-winged, 84
Crotophaga
   sulcirostris, 47
Crow,
   American, 58
   Fish, 58
Cuckoo,
   Black-billed, 47
   Yellow-billed, 47

Curlew,
  Eskimo, 36
  Long-billed, 37
  Slender-billed, 37
Cyanocitta
  cristata, 57
Cygnus
  buccinator, 20
  columbianus, 19
  cygnus, 87
  olor, 20
Cynanthus
  latirostris, 50

Dendragapus
  canadensis, 30, 89–90
Dendrocygna
  bicolor, 19, 88
Dendroica
  caerulescens, 68
  castanea, 70
  cerulea, 70
  coronata, 68
  discolor, 70
  dominica, 69
  fusca, 69
  kirtlandii, 70
  magnolia, 68
  nigrescens, 68
  occidentalis, 69
  palmarum, 70
  pensylvanica, 68
  petechia, 67, 93–94
  pinus, 69
  striata, 70
  tigrina, 68
  townsendi, 69
  virens, 69
Dickcissel, 75
Dolichonyx
  oryzivorus, 80–81
Dove,
  Common Ground-, 47
  Mourning, 46, 91
  Ringed Turtle-, 87
  Rock, 46
  White-winged, 46
Dovekie, 45

Dowitcher,
  Long-billed, 40
  Short-billed, 39–40, 91
Dryocopus
  pileatus, 52
Duck,
  American Black, 6, 21, 88–89
  Fulvous Whistling-, 19, 88
  Harlequin, 24
  Ring-necked, 23
  Ruddy, 26
  Tufted, 23
  Wood, 21
Dumetella
  carolinensis, 63
Dunlin, 39

Eagle,
  Bald, 27, 89
  Golden, 29
Ectopistes
  migratorius, 46–47
Egret,
  Cattle, 18
  Great, 17
  Snowy, 18
Egretta
  caerulea, 18
  thula, 18
  tricolor, 18
Eider,
  Common, 24, 89
  King, 24
Elanoides
  forficatus, 26–27
Empidonax
  alnorum, 53
  difficilis, 54
  flaviventris, 53
  minimus, 54
  traillii, 53
  virescens, 53
  wrightii, 54
Eremophila
  alpestris, 55–56, 92
Eudocimus
  albus, 19
  ruber, 87

Euphagus
 carolinus, 82
 cyanocephalus, 82

Falco
 columbarius, 29, 89
 mexicanus, 30
 peregrinus, 29, 89
 rusticolus, 30
 sparverius, 29
Falcon,
 Peregrine, 29, 89
 Prairie, 30
Fieldfare, 63
Finch,
 House, 84
 Purple, 83–84
 Rosy, 7, 83
Flamingo,
 Greater, 19
Flicker,
 Northern, 7, 52, 92
 Red-shafted, 92
 Yellow-shafted, 92
Flycatcher,
 Acadian, 53
 Alder, 53
 Ash-throated, 54
 Fork-tailed, 55
 Gray, 54
 Great Crested, 54
 Least, 54
 Olive-sided, 53
 Scissor-tailed, 55
 Sulphur-bellied, 55
 Vermilion, 54
 Western, 54
 Willow, 53
 Yellow-bellied, 53
Fratercula
 arctica, 46
Fregata
 magnificens, 17
Frigatebird,
 Magnificent, 17
Fringilla
 montifringilla, 83
Fulica
 americana, 33

Fulmar,
 Northern, 15, 88
Fulmarus
 glacialis, 15, 88

Gadwall, 22
Gallinago
 gallinago, 40
Gallinula
 chloropus, 33
Gallinule,
 Purple, 33
Gannet,
 Northern, 16
Gavia
 adamsii, 14
 immer, 14
 pacifica, 14
 stellata, 14
Geothlypis
 trichas, 72, 94
Gnatcatcher,
 Blue-gray, 61
Godwit,
 Hudsonian, 37
 Marbled, 37
Goldeneye,
 Barrow's, 25
 Common, 25
Golden-Plover,
 Lesser. See Plover, Lesser Golden-
Goldfinch,
 American, 85, 96
 European, 88
 Lesser, 85
Goose,
 Barnacle, 21
 Bean, 87
 Canada, 21, 88
 Greater White-fronted, 20
 Greater Snow, 88
 Ross', 20
 Snow, 7, 20
Goshawk,
 Northern, 27–28
Grackle,
 Common, 82
 Great-tailed, 82

Grebe,
  Eared, 15
  Horned, 14
  Pied-billed, 14
  Red-necked, 15
  Western, 15
Grosbeak,
  Black-headed, 74
  Blue, 74
  Evening, 85
  Pine, 83, 95–96
  Rose-breasted, 74
Ground-Dove,
  Common. See Dove, Common Ground-
Grouse,
  Ruffed, 31, 90
  Sharp-tailed, 31, 90
  Spruce, 30, 89–90
Grus
  americana, 33
  canadensis, 33, 90
Guillemot,
  Black, 12, 45–46, 91
Guiraca
  caerulea, 74
Gull,
  Bonaparte's, 42
  California, 42
  Common Black-headed, 41
  Franklin's, 41
  Glaucous, 43
  Great Black-backed, 43
  Herring, 42
  Iceland, 7, 42, 91
  Ivory, 43
  Laughing, 41
  Lesser Black-backed, 43, 91
  Little, 41
  Mew, 42
  Ring-billed, 42
  Ross', 43
  Sabine's, 43
  Thayer's. See Gull, Iceland
Gyrfalcon, 30

Haematopus
  palliatus, 35

Haliaeetus
  leucocephalus, 27, 89
Harrier,
  Northern, 27
Hawk,
  Broad-winged, 28
  Cooper's, 27
  Red-shouldered, 28
  Red-tailed, 28, 89
  Rough-legged, 28
  Sharp-shinned, 27
  Swainson's, 28
Helmitheros
  vermivorus, 71
Heron,
  Black-crowned Night-, 18
  Great Blue, 17
  Green-backed, 18
  Little Blue, 18
  Tricolored, 18
  Yellow-crowned Night-, 19
Heteroscelus
  incanus, 36
Himantopus
  mexicanus, 35
Hirundo
  fulva, 57
  pyrrhonota, 56–57, 93
  rustica, 57
Histrionicus
  histrionicus, 24
Hummingbird,
  Broad-billed, 50
  Ruby-throated, 51
  Rufous, 51
Hylocichla
  mustelina, 63

Ibis,
  Glossy, 19
  Scarlet, 87
  White, 19
  White-faced, 19
Icteria
  virens, 73
Icterus
  galbula, 83
  parisorum, 83
  spurius, 83

Ictinia
  mississippiensis, 27
Ixobrychus
  exilis, 17
Ixoreus
  naevius, 63

Jackdaw, 57
Jaeger,
  Long-tailed, 41
  Parasitic, 41
  Pomarine, 41
Jay,
  Blue, 57
  Gray, 57
  Scrub, 87
Junco,
  Dark-eyed, 7, 80, 95
  Oregon, 95
Junco
  hyemalis, 80, 95

Kestrel,
  American, 29
Killdeer, 35
Kingbird,
  Cassin's, 55
  Eastern, 55
  Gray, 55
  Western, 55
Kingfisher,
  Belted, 51
Kinglet,
  Golden-crowned, 60–61
  Ruby-crowned, 61
Kite,
  American Swallow-tailed, 26–27
  Mississippi, 27
Kittiwake,
  Black-legged, 43
Knot,
  Red, 37

Lagopus
  lagopus, 30, 90
  mutus, 30–31
Lanius
  excubitor, 65, 93
  ludovicianus, 65

Lark,
  Horned, 55–56, 92
Larus
  argentatus, 42
  atricilla, 41
  californicus, 42
  canus, 42
  delawarensis, 42
  fuscus, 43, 91
  glaucoides, 42, 91
  hyperboreus, 43
  marinus, 43
  minutus, 41
  philadelphia, 42
  pipixcan, 41
  ridibundus, 41
  thayeri. See *Larus glaucoides*
Laterallus
  jamaicensis, 32
Leucosticte
  arctoa, 83
Limnodromus
  griseus, 39–40, 91
  scolopaceus, 40
Limnothlypis
  swainsonii, 71
Limosa
  fedoa, 37
  haemastica, 37
Linnet, 88
Longspur,
  Chestnut-collared, 80
  Lapland, 80
  Smith's, 80
Loon,
  Common, 14
  Pacific, 14
  Red-throated, 14
  Yellow-billed, 14
Lophodytes
  cucullatus, 25
Loxia
  curvirostra, 84, 96
  leucoptera, 84
Luscinia
  calliope, 61

Magpie,
  Black-billed, 57

Mallard, 22
Martin,
  Purple, 56
Meadowlark,
  Eastern, 81
  Western, 81
Melanerpes
  carolinus, 51
  erythrocephalus, 51, 92
  lewis, 51
Melanitta
  fusca, 25
  nigra, 24
  perspicillata, 24
Meleagris
  gallopavo, 31–32
Melospiza
  georgiana, 79
  lincolnii, 79
  melodia, 78, 95
Merganser,
  Common, 25
  Hooded, 25
  Red-breasted, 26
Mergellus
  albellus, 25
Mergus
  merganser, 25
  serrator, 26
Merlin, 29, 89
Mimus
  polyglottos, 63–64
Mniotilta
  varia, 71
Mockingbird,
  Northern, 63–64
Molothrus
  ater, 82, 95
Moorhen,
  Common, 33
Morus
  bassanus, 16
Murre,
  Thick-billed, 45
Murrelet,
  Ancient, 46
Myadestes
  townsendi, 62

Mycteria
  americana, 19
Myiarchus
  cinerascens, 54
  crinitus, 54
Myioborus
  pictus, 73
Myiodynastes
  luteiventris, 55
Myiopsitta
  monachus, 47

Nandayus
  nenday, 87
Nighthawk,
  Common, 50
  Lesser, 50
Night-Heron,
  Black-crowned, 18
  Yellow-crowned, 19
Nucifraga
  columbiana, 57
Numenius
  americanus, 37
  borealis, 36
  phaeopus, 36
  tenuirostris, 37
Nutcracker,
  Clark's, 57
Nuthatch,
  Red-breasted, 59
  White-breasted, 59
Nyctanassa
  violacea, 19
Nyctea
  scandiaca, 48
Nycticorax
  nycticorax, 18

Oceanites
  oceanicus, 15
Oceanodroma
  castro, 16
  leucorhoa, 16
Oenanthe
  oenanthe, 61
Oldsquaw, 24
Oporornis
  agilis, 72

formosus, 72
philadelphia, 72
tolmiei, 72, 94
Oreoscoptes
  montanus, 64
Oriole,
  Northern, 7, 83
  Orchard, 83
  Scott's, 83
Osprey, 26
Otus
  asio, 48
Ovenbird, 71, 94
Owl,
  Barn, 7, 48
  Barred, 49
  Boreal, 49
  Burrowing, 48
  Eastern Screech-, 48
  Great Gray, 49
  Great Horned, 48, 91–92
  Long-eared, 49
  Northern Hawk, 48
  Northern Saw-whet, 49
  Short-eared, 49
  Snowy, 48
Oxyura
  jamaicensis, 26
Oystercatcher,
  American, 35

Pagophila
  eburnea, 43
Pandion
  haliaetus, 26
Parakeet,
  Black-hooded, 87
  Carolina, 47
  Monk, 47
Partridge,
  Gray, 30, 89
Parula,
  Northern, 67
Parula
  americana, 67
Parus
  atricapillus, 58
  bicolor, 59
  caeruleus, 87

carolinensis, 58, 93
gambeli, 58
hudsonicus, 58–59
Passer
  domesticus, 85
Passerculus
  sandwichensis, 77, 95
Passerella
  iliaca, 78
Passerina
  amoena, 74
  ciris, 75
  cyanea, 74
Pelecanus
  erythrorhynchos, 16
  occidentalis, 16
Pelican,
  American White, 16
  Brown, 16
Perdix
  perdix, 30, 89
Perisoreus
  canadensis, 57
Petrel,
  Band-rumped Storm-, 16
  Black-capped, 15
  Leach's Storm-, 16
  Wilson's Storm-, 15
Pewee,
  Eastern Wood-, 53
  Western Wood-, 53
Phainopepla, 64
Phainopepla
  nitens, 64
Phalacrocorax
  auritus, 16
  carbo, 16
Phalaenoptilus
  nuttallii, 50
Phalarope,
  Red, 40
  Red-necked, 40
  Wilson's, 40
Phalaropus
  fulicaria, 40
  lobatus, 40
  tricolor, 40
Phasianus
  colchicus, 30, 89

Pheasant,
  Ring-necked, 30, 89
Pheucticus
  ludovicianus, 74
  melanocephalus, 74
Philomachus
  pugnax, 39
Phoebe,
  Eastern, 54
  Say's, 54
Phoenicopterus
  ruber, 19
Pica
  pica, 57
Picoides
  arcticus, 52
  pubescens, 52, 92
  tridactylus, 52
  villosus, 52, 92
Pigeon,
  Band-tailed, 46, 91
  Passenger, 46–47
Pinicola
  enucleator, 83, 95–96
Pintail,
  Northern, 22
  White-cheeked, 87
Pipilo
  chlorurus, 75
  erythrophthalmus, 75, 94
Pipit,
  American, 7, 64
  Sprague's, 64
Piranga
  ludoviciana, 74
  olivacea, 73
  rubra, 73
Plectrophenax
  nivalis, 80
Plegadis
  chihi, 19
  falcinellus, 19
Plover,
  Black-bellied, 34
  Lesser Golden-, 34
  Mongolian, 34
  Piping, 34, 90
  Semipalmated, 34
  Snowy, 34
  Wilson's, 34
Pluvialis
  dominica, 34
  squatarola, 34
Podiceps
  auritus, 14
  grisegena, 15
  nigricollis, 15
Podilymbus
  podiceps, 14
Polioptila
  caerulea, 61
Polyborus
  plancus, 29
Pooecetes
  gramineus, 76–77, 94
Poor-will,
  Common, 50
Porphyrula
  martinica, 33
Porzana
  carolina, 32
Prairie-Chicken,
  Greater, 31
Progne
  subis, 56
Protonotaria
  citrea, 71
Ptarmigan,
  Rock, 30–31
  Willow, 30, 90
Pterodroma
  hasitata, 15
Puffin,
  Atlantic, 46
Puffinus
  lherminieri, 15
Pyrocephalus
  rubinus, 54
Pyrrhuloxia, 87

Quiscalus
  mexicanus, 82
  quiscula, 82

Rail,
  Black, 32
  King, 32
  Virginia, 32

Yellow, 32
Rallus
  elegans, 32
  limicola, 32
Raven,
  Chihuahuan, 87
  Common, 58
Razorbill, 45
Recurvirostra
  americana, 35
Redhead, 23
Redpoll,
  Common, 84, 96
  Hoary, 6, 85, 96
Redshank,
  Spotted, 35
Redstart,
  American, 71, 94
  Painted, 73
Regulus
  calendula, 61
  satrapa, 60–61
Rhodostethia
  rosea, 43
Riparia
  riparia, 56
Rissa
  tridactyla, 43
Robin,
  American, 63
Rubythroat,
  Siberian, 61
Ruff, 39
Rynchops
  niger, 45

Salpinctes
  obsoletus, 59
Sanderling, 37
Sandpiper,
  Baird's, 38
  Buff-breasted, 39
  Curlew, 39
  Least, 38
  Pectoral, 38
  Purple, 39
  Semipalmated, 37–38
  Sharp-tailed, 38
  Solitary, 36

Spotted, 36
Stilt, 39
Upland, 36
Western, 38
White-rumped, 38
Sapsucker,
  Yellow-bellied, 51
Sayornis
  phoebe, 54
  saya, 54
Scaup,
  Greater, 23
  Lesser, 24
Scolopax
  minor, 40
Scoter,
  Black, 24
  Surf, 24
  White-winged, 25
Screech-Owl,
  Eastern. See Owl, Eastern Screech-
Seiurus
  aurocapillus, 71, 94
  motacilla, 72
  noveboracensis, 71, 94
Selasphorus
  rufus, 51
Serin, 87
Serinus
  serinus, 87
Setophaga
  ruticilla, 71, 94
Shearwater,
  Audubon's, 15
Shelduck,
  Ruddy, 87
Shoveler,
  Northern, 22
Shrike,
  Loggerhead, 65
  Northern, 65, 93
Sialia
  currucoides, 61–62
  sialis, 61
Siskin,
  Eurasian, 88
  Pine, 85
Sitta
  canadensis, 59

carolinensis, 59
Skimmer,
   Black, 45
Skua,
   Great, 41
Smew, 25
Snipe,
   Common, 40
Solitaire,
   Townsend's, 62
Somateria
   mollissima, 24, 89
   spectabilis, 24
Sora, 32
Sparrow,
   American Tree, 76
   Bachman's, 75
   Baird's, 77
   Brewer's, 76
   Cassin's, 75
   Chipping, 76, 94
   Clay-colored, 76
   Field, 76
   Fox, 78
   Golden-crowned, 79
   Grasshopper, 77, 95
   Harris', 80
   Henslow's, 78
   House, 85
   Lark, 77
   Le Conte's, 78
   Lincoln's, 79
   Savannah, 77, 95
   Sharp-tailed, 78, 95
   Song, 78, 95
   Swamp, 79
   Vesper, 76–77, 94
   White-crowned, 79, 95
   White-throated, 79
Sphyrapicus
   varius, 51
Spiza
   americana, 75
Spizella
   arborea, 76
   breweri, 76
   pallida, 76
   passerina, 76, 94
   pusilla, 76

Starling,
   European, 65
Stelgidopteryx
   serripennis, 56
Stercorarius
   longicaudus, 41
   parasiticus, 41
   pomarinus, 41
Sterna
   antillarum, 45
   caspia, 44
   dougallii, 44
   forsteri, 44
   fuscata, 45
   hirundo, 44
   maxima, 44
   paradisaea, 44
   sandvicensis, 44
Stilt,
   Black-necked, 35
Stint,
   Little, 38
   Rufous-necked, 38
Stork,
   Wood, 19
Storm-Petrel,
   Band-rumped, 16
   Leach's, 16
   Wilson's, 15
Streptopelia
   risoria, 87
Strix
   nebulosa, 49
   varia, 49
Sturnella
   magna, 81
   neglecta, 81
Sturnus
   vulgaris, 65
Surnia
   ulula, 48
Swallow,
   Bank, 56
   Barn, 57
   Cave, 57
   Cliff, 56–57, 93
   Northern Rough-winged, 7, 56
   Tree, 56

Swan,
 Mute, 20
 Trumpeter, 20
 Tundra, 7, 19
 Whooper, 87
Swift,
 Chimney, 50
Synthliboramphus
 antiquus, 46

Tachycineta
 bicolor, 56
Tadorna
 ferruginea, 87
Tanager,
 Scarlet, 73
 Summer, 73
 Western, 74
Tattler,
 Wandering, 36
Teal,
 Blue-winged, 22, 89
 Cinnamon, 22
 Green-winged, 7, 21
Tern,
 Arctic, 44
 Black, 45
 Caspian, 44
 Common, 44
 Forster's, 44
 Least, 45
 Roseate, 44
 Royal, 44
 Sandwich, 44
 Sooty, 45
Thrasher,
 Brown, 64, 93
 Sage, 64
Thrush,
 Gray-cheeked, 62, 93
 Hermit, 62
 Swainson's, 62
 Varied, 63
 Wood, 63
Thryomanes
 bewickii, 60, 93
Thryothorus
 ludovicianus, 59

Tit,
 Blue, 87
Titmouse,
 Tufted, 7, 59
Towhee,
 Green-tailed, 75
 Rufous-sided, 75, 94
Toxostoma
 rufum, 64, 93
Tringa
 erythropus, 35
 flavipes, 35
 melanoleuca, 35
 solitaria, 36
Troglodytes
 aedon, 60, 93
 troglodytes, 60
Tryngites
 subruficollis, 39
Turdus
 merula, 63
 migratorius, 63
 pilaris, 63
Turkey,
 Wild, 31–32
Turnstone,
 Ruddy, 37, 90
Turtle-Dove,
 Ringed. See Dove, Ringed Turtle-
Tympanuchus
 cupido, 31
 phasianellus, 31, 90
Tyrannus
 dominicensis, 55
 forficatus, 55
 savana, 55
 tyrannus, 55
 verticalis, 55
 vociferans, 55
Tyto
 alba, 48

Uria
 lomvia, 45

Veery, 62, 93
Vermivora
 celata, 67
 chrysoptera, 66–67

peregrina, 67
pinus, 66
ruficapilla, 67
virginiae, 67
Vireo,
　Bell's, 65
　Philadelphia, 66
　Red-eyed, 66
　Solitary, 65–66
　Warbling, 66
　White-eyed, 65
　Yellow-throated, 66
Vireo
　bellii, 65
　flavifrons, 66
　gilvus, 66
　griseus, 65
　olivaceus, 66
　philadelphicus, 66
　solitarius, 65–66
Vulture,
　Black, 26
　Turkey, 26

Warbler,
　Bay-breasted, 70
　Black-and-white, 71
　Blackburnian, 69
　Blackpoll, 70
　Black-throated Blue, 68
　Black-throated Gray, 68
　Black-throated Green, 69
　Blue-winged, 66
　Brewster's, 7, 66
　Canada, 73
　Cape May, 68
　Cerulean, 70
　Chestnut-sided, 68
　Connecticut, 72
　Golden-winged, 66–67
　Hermit, 69
　Hooded, 72
　Kentucky, 72
　Kirtland's, 70
　Lawrence's, 7, 66
　MacGillivray's, 72, 94
　Magnolia, 68
　Mourning, 72
　Nashville, 67
　Orange-crowned, 67
　Palm, 7, 70
　Pine, 69
　Prairie, 70
　Prothonotary, 71
　Swainson's, 71
　Tennessee, 67
　Townsend's, 69
　Virginia's, 67
　Wilson's, 73
　Worm-eating, 71
　Yellow, 67, 93–94
　Yellow-rumped, 7, 68
　Yellow-throated, 69
Waterthrush,
　Louisiana, 72
　Northern, 71, 94
Waxwing,
　Bohemian, 64
　Cedar, 64
Wheatear,
　Northern, 61
Whimbrel, 36
Whip-poor-will, 50
Whistling-Duck,
　Fulvous, 19, 88
Wigeon,
　American, 23
　Eurasian, 22
Willet, 36
Wilsonia
　canadensis, 73
　citrina, 72
　pusilla, 73
Woodcock,
　American, 40
Woodpecker,
　Black-backed, 52
　Downy, 52, 92
　Hairy, 52, 92
　Lewis', 51
　Pileated, 52
　Red-bellied, 51
　Red-headed, 51, 92
　Three-toed, 52
Wood-Pewee,
　Eastern, 53
　Western, 53

Wren,
  Bewick's, 60, 93
  Carolina, 59
  House, 60, 93
  Marsh, 60
  Rock, 59
  Sedge, 60
  Winter, 60

Xanthocephalus
  xanthocephalus, 81
Xema
  sabini, 43

Yellowlegs,
  Greater, 35
  Lesser, 35
Yellowthroat,
  Common, 72, 94

Zenaida
  asiatica, 46
  macroura, 46, 91
Zonotrichia
  albicollis, 79
  atricapilla, 79
  leucophrys, 79, 95
  querula, 80